D0219498

Ageing and Youth Culture

DISCARD

Ageing and Youth Culture
Music, Style and Identity

Edited by
Andy Bennett and Paul Hodkinson

London • New York

English edition
First published in 2012 by
Berg
Editorial offices:
50 Bedford Square, London WC1B 3DP, UK
175 Fifth Avenue, New York, NY 10010, USA

© Andy Bennett and Paul Hodkinson 2012

All rights reserved.
No part of this publication may be reproduced in any form
or by any means without the written permission of Berg.

Berg is an imprint of Bloomsbury Publishing Plc.

Library of Congress Cataloging-in-Publication Data

A catalogue record for this book is available from the Library of Congress.

British Library Cataloguing-in-Publication Data

A catalogue record for this book is available from the British Library.

ISBN 978 1 84788 836 5 (Cloth)
978 1 84788 835 8 (Paper)
e-ISBN 978 0 85785 295 3 (ePDF)
978 0 85785 037 9 (epub)

Typeset by Apex CoVantage, LLC, Madison, WI, USA.
Printed in the UK by the MPG Books Group

www.bergpublishers.com

Contents

About the Authors

Andy Bennett is Professor of Cultural Sociology and director of the Griffith Centre for Cultural Research at Griffith University in Queensland, Australia. He has authored and edited numerous books, including *Popular Music and Youth Culture* (Macmillan, 2000), *Cultures of Popular Music* (Open University Press, 2001), *After Subculture* (Palgrave, 2004), *Remembering Woodstock* (Ashgate, 2004) and *Music Scenes* (with Richard A. Peterson, Vanderbilt University Press, 2004). He is currently lead Chief Investigator on a three-year, five-country project funded by the Australian Research Council entitled 'Popular Music and Cultural Memory: Localized Popular Music Histories and Their Significance for National Music Industries'. He is editor-in-chief of the *Journal of Sociology*, a faculty fellow of the Center for Cultural Sociology, Yale University, and an associate member of PopuLUs, the Centre for the Study of the World's Popular Musics, Leeds University.

Joanna R. Davis completed her PhD in Sociology at the University of California, Santa Barbara, where her dissertation examined music, identity and the transition to adulthood in a local punk scene. Her interest in the intersection between identity and cultural discourses has also led to publications on such topics as breastfeeding in public and gender in popular culture. Recently, she has turned her attention to youth development research and programming at nonprofit organizations. She currently resides in Eugene, Oregon.

Mary Fogarty is Assistant Professor in Dance at York University (Toronto, Canada). She is the author of 'A Manifesto for the Study of Popular Dance' (2010) in *Conversations Across the Field of Dance Studies* (Society of Dance History Scholars) and various articles about aspects of breaking, waacking and other hip-hop, funk and disco dances. She is a practising b-girl/ethnographer and continues to teach practical workshops for choreographers, dancers, youth and prisoners, with occasional performances on theatre stages and in competitions. She completed her PhD in Music at the University of Edinburgh (2011), supervised by Simon Frith.

Lucy Gibson was a Lecturer in Sociology at the University of Manchester from 2008 to 2011. Her PhD research explored popular music and the life course, and investigated themes of cultural commitment, lifestyles and identities amongst fans of Northern Soul, rock and electronic dance music. She has published work on popular music and ageing, using email interviews to investigate music and memory and popular music more generally. She is currently working as an independent scholar.

Danielle Giffort is a graduate student in Sociology at the University of Illinois at Chicago. Her research interests include gender, organizations, subcultures and social movements. Her work has appeared in *Gender and Society*. She is currently working on a research project that explores women's participation in the marijuana legalization movement.

Julie Gregory is a PhD candidate in Sociology at Queen's University in Kingston, Ontario, Canada. Her dominant research interests concern the sociology of knowledge production and circulation, with special attention paid to intersecting power relations that support (or not) the emergence and proliferation of 'problem-solutions'. For example, in a recent article published in *Social and Legal Studies*, she problematizes one community response to drug use by situating it in the context of dominant discourses that construct drug use and parenthood—particularly motherhood—as necessarily incongruous.

Ross Haenfler is an Associate Professor of Sociology at the University of Mississippi. He is the author of *Straight Edge: Clean Living Youth, Hardcore Punk, and Social Change* (Rutgers, 2006) and *Goths, Gamers, and Grrrls: Deviance and Youth Subcultures* (Oxford, 2010). Ross has published a variety of articles in the areas of social movements, subcultures and gender. His current projects include studies of lifestyle movements such as voluntary simplicity and virginity pledgers, and how participants in youth cultures transition to adulthood. An award-winning teacher, his courses include social movements, deviance and youth subcultures, men and masculinities, and political sociology.

Paul Hodkinson is Senior Lecturer in Sociology at the University of Surrey. He is the author of *Goth: Identity, Style and Subculture* (Berg, 2002) and a variety of articles and chapters relating to questions of youth, subculture, identity and media. He is also co-editor of *Youth Cultures: Scenes, Subcultures and Tribes* (Routledge, 2007) and the author of *Media, Culture and Society* (Sage, 2010). He is co-editor of the journal *Sociological Research Online*. His most recent research and writing has focused on questions of ageing and identity amongst older goths.

Samantha Holland is a Research Fellow at Leeds Metropolitan University, where her work centres mostly around gender, ageing and nonmainstream leisure and subcultures. Previous publications include *Alternative Femininities: Body, Age and Identity* (Berg, 2004), *Remote Relationships in a Small World* (editor, Peter Lang, 2008) and *Pole Dancing, Empowerment and Embodiment* (Palgrave Macmillan, 2010). She is currently writing a book about women's roller derby titled *'Take that*

Bitch Down!' Gendered Violence and Embodiment in Women's Roller Derby (SUNY Press, 2012).

Kristen Schilt is an Assistant Professor of Sociology at the University of Chicago. She is the author of *Just One of the Guys: Transgender Men and the Persistence of Gender Inequality* (University of Chicago Press, 2010), as well as a variety of articles about feminist subcultures. Her research centres on gender inequality, embodiment, subculture and the workplace.

Nicola Smith is Senior Lecturer in Sociology and Contemporary Media at Cardiff Metropolitan University. Her research interests include the British Northern Soul scene, adult-frequented and ageing music cultures, post-subcultural theory and the performance of identities within fandom. Recent publications include 'Beyond the Master Narrative of Youth: Ageing Popular Music Scenes' (2009) in *The Ashgate Research Companion to Popular Musicology* (Ashgate).

Jodie Taylor is a Postdoctoral Research Fellow at the Griffith Centre for Cultural Research, Griffith University. She is the author of *Playing it Queer: Popular Music, Identity and Queer World-Making* (Peter Lang, 2012) and co-editor of two forthcoming collections, *The Festivalisation of Culture: Place, Identity and Politics* (Ashgate, 2012) and *Redefining Mainstream Popular Music* (Routledge, 2012). She is also co-editor of a special edition of *Continuum* on 'Erotic Screen and Sound'. Additionally, Jodie has authored numerous journal articles and book chapters on queer scenes, popular music and identity, and ethnographic methods.

Bill Tsitsos is an Assistant Professor in the department of Sociology, Anthropology, and Criminal Justice at Towson University. His research reflects interests in both the sociology of religion and the sociology of music-based subcultures. His 1999 article 'Rules of Rebellion: Slamdancing, Moshing, and the American Alternative Scene' examined dancing among younger members of the punk scene. In this volume, he returns to that topic, with a focus on older scene members. He is currently finishing a paper that compares white musicians who claim to be aliens from outer space with nonwhite alien musicians, using racial transparency theory.

Introduction

Andy Bennett and Paul Hodkinson

In the early twenty-first century, the concept of 'youth culture' appears increasingly ambiguous and open to interpretation. Born out of a unique combination of socioeconomic growth and rapid technological development during the 1950s, and spearheaded by the emergence of the 'affluent' teenager, it would be fair to say that 'youth culture' has always denoted a marketing strategy as much as it has a lifestyle (Osgerby 2001). Obviously, this is not to categorize youth culture in any absolute sense as a commercially created venture. Indeed, a significant volume of work, from the Centre for Contemporary Cultural Studies (CCCS), subcultural theory of the 1970s (Hall and Jefferson 1976) through to more recent work grounded in post-subculturalist (Bennett 2000; Muggleton 2000) and consumerist (Miles 2000; Cummings 2008) perspectives has sought to illustrate—albeit, from somewhat differing theoretical positions—the ways in which images, objects and texts produced by the youth market are appropriated symbolically by adolescents and become grounded in their everyday reality. From the point of view of the CCCS, such appropriation was bound up with the coming together of issues of class inequality and disempowerment and those of emerging adolescence—with these being felt most keenly by working-class youth whose spectacular stylistic assemblages were theorized as youthful rituals of resistance (Clarke et al. 1976) and semiotic guerrilla warfare (Hebdige 1979). Post-subcultural theory, although less focused around class, has tended in its exploration of postmodern consumerism and the reflexive construction of identity, to share with the CCCS approach a presentation of youth culture as an age-specific and age-limited category confined in most understandings to the teens and early twenties (Muggleton and Wienzierl 2003).

Interestingly, this interpretation is out of step with research emerging from the field of youth transitions, in which a range of theorists have drawn attention to the apparently increasing diversity, complexity and longevity of youth and the porous nature of the boundaries between adolescence and adulthood. This is reflected in the increasing use of terms such as 'emerging adulthood' (Tanner and Arnett 2009) and 'young adulthood' (Furlong 2009) in order to make sense of the increasing prevalence—particularly among middle-class youth—of extended transitions in which firm adult commitments such as marriage, childbearing, home ownership and dedication to career are taken on more gradually or unevenly than in the past, while leisure and lifestyle habits remain 'youthful' for longer (Du Bois-Reymond 1998).

Consistent with this broader context, while everyday responses to new musical genres and associated forms of cultural practice in the 1950s, 1960s and 1970s could be aligned with youth in a narrow sense (e.g. see Bradley 1992; Hebdige 1979; Roszak 1969), in more recent times the straightforward equation of youth cultures with the young has become more difficult to sustain. Many of those musically and stylistically based groupings once unproblematically referred to as *youth* cultures are now increasingly multigenerational. Indeed, the continuing participation of older generations is such that it often goes beyond the delayed or uneven youth transitions alluded to by concepts such as 'emerging' or 'young' adulthood. Such ongoing involvement does not always reflect a delay or refusal of adult commitments; nor is it confined to a limited period of early or emerging adulthood. Not only are individuals continuing to participate in punk, goth, bebop, Northern Soul, rock and dance cultures, as well as various other music scenes, into and beyond their thirties, but many are also doing so alongside substantial career, family and other commitments, and the development of other familiar facets of adulthood and ageing.

Such a development does not necessarily signify an 'end of youth culture' in any qualitative sense. Notably, the sphere of popular music—together with associated stylistic innovations—is characterized by a continuing progression of new genres and subgenres that are primarily marketed to, consumed by and developed among the young. Nevertheless, multigenerational investment in an increasing array of popular music and related styles originating in the latter part of the twentieth century does have a significant impact on the concept of youth culture.

There has been some limited engagement with this issue already. For example, Weinstein's (2000) study of heavy metal refers to ageing fans of the genre as 'wistful emigrants', no longer able to fully participate in the scene, with involvement restricted to occasionally playing their old records (2000: 111). Similarly, in her study of the US punk scene, Andes (1998) argues that for ageing members of this scene, participation typically becomes restricted to roles in areas such as production and promotion. In contrast to Andes's findings, Bennett's (2006) study of ageing punks in the United Kingdom suggests a less restricted series of options for scene participation, with those in their forties and fifties still attending, and sometimes performing at, gigs. Similarly, Bennett's work also reveals how age can offer avenues for the articulation of 'status' garnered through experience and longevity as a 'fully paid up' punk. Across a range of other 'youth' cultures, one sees evidence of such traits as older participants continue to *buy into*, both in a literal and an aesthetic sense, particular music and styles in which they have been culturally invested for many years (e.g. see Holland 2004; Haenfler 2006; Davies 2006; Gregory 2009; Smith 2009; Taylor 2010; Hodkinson 2011).

For some commentators, this presents a sociocultural conundrum—evidence of a social pathology in which individuals are refusing to grow up and attempting to hold on to their youth through whatever means are necessary (Calcutt 1998). In another sense, however, it is possible to see how such a continuation in taste patterns among ageing individuals is symbolic of the transformation of consumer lifestyles into modes of cultural empowerment through which ageing individuals continue to construct and articulate identities and claim distinctiveness in contemporary everyday life. Such interpretations of the significance of music-related affiliations and practices have been offered regularly in relation to younger fans, but much less attention has been given to the continuing influence of such factors in a 'post-youth' context. However, as Bennett observes:

> One might reasonably expect . . . that where investment in a musical style has been particularly intensive during one's teenage or twenty-something years such investment may continue past thirty, into middle age and perhaps later life. (2006: 221)

The same, of course, can be applied to aspects of visual style, and other cultural resources and accessories that align with musical taste.

To explore this observation a little further, there may be a case here for a recasting of youth culture, not merely as tied to an age-specific period of transition in the life course. Rather, by dint of its embeddedness in the consumer-based fabric of contemporary everyday life, youth culture has taken on a more malleable property. While at one level it continues to bespeak the cultural expression of a social demographic in transition from childhood to adulthood, at another level certain key elements of youth culture have expanded and extended in ways that increasingly have become more connected and compatible with adult lives.

These shifting parameters of youthfulness and adulthood correspond with shifts in the industrial production and representation of music, style and associated cultural resources. No longer marketed purely as objects of youth consumption, some of these products increasingly are targeted at a much broader demographic. A number of studies have already noted the palpable effects of this in relation to ageing articulations of taste in music, fashion and style once deemed to be restricted to the cultural repertoire of youth (see Blaikie 1999; Holland 2004; Bennett 2009; Hodkinson 2011). Importantly, such work builds on previous observations made in research on ageing and the life course that have noted how shifting definitions of ageing are giving rise to new expectations, priorities and understandings among those in middle age and, increasingly, later life. For example, as Hunt observes:

> These are people who know that they have another 30–40 years of life expectancy ahead of them. They may see a practically endless future rather than the beginning of the end. (2005: 183)

That such shifts are forcing new understandings of youth culture and of people's trajectories through different forms of adulthood is clear. Less clear, however, are the detailed ways in which such shifts are negotiated on an individual and collective level within different youth cultures, or indeed the ways in which these negotiations are informed by expectations and constraints relating to socioeconomic background or status. Critical here is an understanding of how the identities and lifestyles constructed by 'post-youth' individuals may often need to include the accommodation of new demands, expectations and compromises created, for example, because of work and family commitments that invariably accompany the transition from early to middle adulthood. They may also involve adaptations and developments in participation as a result of ageing bodies whose perceived or actual compatibility with the most overtly youth-oriented images, styles and practices may recede sharply as a result of the ageing process. Continuing participation also develops and responds to broader changes in priorities, outlooks and self-perceptions that materialize as individuals age.

The purpose of this book is to examine and engage with such issues and questions, and in so doing to begin to set out and define the study of older participation in 'youth' music and style cultures as a key emerging area of study. Drawing on new empirical data from research conducted in the United Kingdom, the United States, Canada and Australia, the chapters that make up the book explore a range of themes connected with ageing in relation to a variety of music scenes and subcultures. The chapters are organized into four themes: ageing, image and identity; constraints of the ageing body; resources and responsibilities; ageing communities. However, readers will note that there are numerous connections between these sections, as well as within them.

The chapters in Part I, Ageing, Image and Identity, focus particularly closely on the development of practices and meanings relating to style and appearance as participants become older. Taking straight edge, a clean-living offshoot of the punk scene, as its case study, Ross Haenfler's chapter examines the way in which, for many of his older respondents, displaying the specificities of spectacular style had gradually become less central to their identities, which now hinged on what was perceived as a broader straight-edge attitude or sensibility. Haenfler explores the ways in which, as their identities have developed into adulthood, individuals have adapted their appearances, avoiding some of the most overt or extreme straight-edge symbols at the same time as retaining telltale elements of subcultural distinctiveness.

While Haenfler's focus is upon the 'toning down' of some of the more overt symbols of youth cultural participation, Jodie Taylor places emphasis on the longevity of transgressive styles and practices among ageing members of Brisbane's queer scene. For Taylor, continued participation in a scene in which music and unconventional appearance connect to queer sexual identities represents a potent challenge to heteronormative understandings of adulthood and the life course, as well as a rejection of dominant models of age and ageing in gay communities.

Julie Gregory's chapter moves the focus from continuing to former participants by exploring the narratives and identities of a group of women who identified as ex-ravers in the Toronto dance music scene. With a particular focus upon the relationship between gender, body image and adulthood, Gregory identifies an emphasis in the accounts of the women on a perceived incompatibility between active participation in rave culture and adult female bodies and responsibilities. She notes, however, that the women all felt that their participation in rave had strongly informed elements of their adult lives.

In Part II, Constraints of the Ageing Body, the emphasis moves to the question of how participants reconcile continuing 'youth' cultural participation with physical limitations of the ageing body. The section begins with an examination of age differences in the competition-centred 'b-boy', or breakdancing, scene as this manifests itself across North America and Western Europe. Focusing on physical questions about the capacity of individuals to continue to dance competitively in their thirties and forties, as well as the dissemination of dance skills from older to young participants, Mary Fogarty suggests that older b-boys tend to develop different roles within the scene, including the coaching of younger members, as the capacity of their own bodies to endure competitive dance reduces. Dance is also the focus of Bill Tsitsos's chapter, which examines the discourses and practices of older punk participants in relation to slam dancing and moshing. Although he identifies a tendency on the part of his older respondents to adopt roles within the middle or rear zones of live music venues, Tsitsos shows how occasional returns to 'the pit' at the front play an important role in the retention or replenishing of fans' physical and emotional connection to the community. Lucy Gibson's chapter takes up the emphasis on dance in the two preceding chapters and combines it with a broader examination of the behaviour of rock fans at live music events in the United Kingdom. As well as noting an increasing tendency to stand still or sit down during performances, Gibson examines practices and sensibilities relating to intoxication, and identifies a broader imperative among many to participate with lower intensity due to bodies that tire more easily and that need to fulfil daytime adult responsibilities.

The orientation of Part III, Resources and Responsibilities, is towards the ways in which participants negotiate continuing involvement in youth cultures with the developing responsibilities and resource issues of developing adult lives. Andy Bennett opens the section with a discussion of his research on dance parties, where ageing participants employ a range of discourses and strategies both to position themselves as credible within the dance party scene and to negotiate the physical demands this places on the ageing body. Like Gibson, Bennett identifies a clear trend among ageing dance party participants towards the implementation of personal safeguards, necessary for balancing their dance party involvement with work and family responsibilities, and referred to by one interviewee as 'sustainable fun'.

Meanwhile, Joanna R. Davis's chapter focuses on the ways in which older punks reconcile an ongoing sense of punk transgression and resistance with everyday adult 'inevitabilities', including careers, long-term relationships and children. Centred on an analysis of the testimonies of key individuals in the scene, Davis highlights the role of accommodation, challenge and redefinition in these older punks' responses to normative expectations of adulthood and their adaption of what it is to be punk.

With a particular focus on gender, Samantha Holland's chapter ends the section by examining the interplay between adult responsibilities and questions of body and image in the development of alternative female identities among former punks, goths and others into their forties and fifties. Revisiting a group of respondents first interviewed over ten years earlier, Holland explores how their gender identities and lifestyles have developed during this period in light of their own ageing and the development of society itself since the 1990s.

While many of the chapters in the first three sections of the book focus on ageing individuals within scenes still dominated by the young, Part IV, Ageing Communities, specifically examines what happens when cohorts of individuals retain and adapt their participation together, or when they begin to pass on their identities to a new generation. Taking the Whitby Gothic Weekend festival as his case study, Paul Hodkinson discusses the collective ageing of a cohort of goths. From topics of conversation to predominant modes of behaviour, and even the format of the event itself, Hodkinson argues that the milieu and orientation of the festival have changed in order to accommodate the developing requirements and priorities of an ageing clientele.

In the final two chapters, emphasis is placed on the transfer of youth cultural skills, knowledge and capital to young people by older generations of continuing participants. Kristen Schilt and Danielle Giffort examine the involvement of former Riot Grrrls in organizing rock 'n' roll camps for girls, which seek to empower young women to become involved in playing music and enable them to challenge female subordination in the rock industry. Schilt and Giffort argue that the camps constitute a development and continuation of Riot Grrl identities and ideologies for the organizers themselves, while creating a space for the emergence of a new generation.

While Schilt and Giffort focus on intergenerational transfer within formal public settings, Nicola Smith's chapter examines the transfer of cultural taste, knowledge and capital within the family. Drawing on her research of the long-standing, multigenerational Northern Soul scene in the United Kingdom, Smith examines the ways in which parents transfer their scene-related affection and expertise to their children, and the differing relationships and responses this gives rise to among 'soul children'.

In summary, the chapters presented here begin to address questions of ageing and youth cultural participation in relation to a variety of specific questions, using case studies from across the spectrum of style and music and from a range of countries across the globe. Collectively, the chapters produce a snapshot of a new 'post-youth' cultural territory that is expanding rapidly to encompass a range of lifestyle and aesthetic sensibilities through which ageing individuals retain tangible cultural connections to tastes and affiliations acquired during their teens and early twenties. While each chapter has its own emphasis, objects and context, as a combined body of work the chapters also begin to develop a range of common themes, questions and areas for further inquiry. The coverage here remains far from exhaustive, but we hope that by bringing together such case studies and developing such themes, this book will help establish and develop the empirical study of ageing 'youth' cultural participants and encourage engagement with the conceptual questions raised by such work about the significance of youth, adulthood and ageing in contemporary societies.

Part I
Ageing, Image and Identity

–1–

'More than the Xs on My Hands': Older Straight Edgers and the Meaning of Style

Ross Haenfler

Introduction

Many scholars have noted the 'spectacular' element of subcultural styles—goths' sinister, macabre fashions (Hodkinson 2002); punks' sculpted hair and garish makeup (Henry 1989); skinheads' boots, braces and shaved heads (Hebdige 1979). For scholars in the Birmingham School tradition, style was an element of semiotic warfare, a symbolic act of resistance among working-class youth to both inequality and 'mainstream' society. Youth associated with music scenes use subcultural styles to help foster a collective identity, to establish authenticity and to set themselves apart from other youth scenes. Less commonly explored questions relate to what happens when spectacular styles meet the demands of adulthood. If articulating an embodied style as a visible, symbolic rejection of dominant social norms and values is at the heart of participation in many music scenes, how do scenesters' music identities and styles change as they age and spectacular fashions become undesirable or inconvenient? If many young people use style to set themselves apart from their peers and 'adult' society, how do the meanings of style change as *they* become adults?

This chapter explores how older adherents of straight edge (sXe)—a clean-living youth scene associated with hardcore punk music—interpret and display their straight-edge affiliation as they age. Straight edgers make a lifetime commitment to abstain from alcohol, tobacco, illegal drugs and often 'casual' sex, communicating their commitment via tattoos and clothing marked with the straight edge symbol, an 'X' (Haenfler 2006; Wood 2006). I draw upon eight years of ethnographic field research and thirty-seven interviews conducted between 1996 and 2004 in Denver, Colorado, and ten follow-up interviews with straight edgers aged over 30 undertaken in 2010. During my fieldwork, I attended over a hundred hardcore shows and socialized with straight-edge kids in a variety of settings. Most of my participants were aged between 17 and 25 (though I also interviewed several 30-somethings). For this current project, I conducted interviews with one woman and nine men, now all over 30 and still straight edge. Nine had earned a university degree (three held postgraduate degrees) and all were employed full time. Their careers included geographer, biologist, community planner, librarian and military officer. Three were married, two were engaged, one had a child and the rest were in committed relationships. I analyzed the interviews in order to search for emergent themes, and compared participants' responses with their transcripts and behaviours from the past.

This chapter shows that for most older participants, straight edge becomes less of an embodied stylistic display and more of a personal philosophy or expression of lifestyle politics. Yet periodic and strategic displays of straight-edge affiliation are still meaningful, communicating longevity in the scene, setting an example for younger adherents or signifying continued resistance to conventional norms. The chapter concludes with a discussion of how adherents of many 'youth' scenes transform the meaning of style as they age, modifying and reframing the visible symbols of youth.

Straight Edge Origins and Styles

Straight edge emerged as an offshoot of the US punk scene, specifically from Washington, DC, hardcore band Minor Threat, whose 1981 song 'Straight Edge' provided the movement with its name. Adherents appreciated the 'question everything' and DIY ethos of punk, but abhorred what they viewed as the scene's self-destructive, nihilist drug and alcohol abuse. Scrawling black Xs on their hands (Xing up)—appropriated from club owners who marked underage kids' hands—straight edgers made it cool *not* to drink. By the mid-1980s, straight-edge scenes had emerged across the United States, and a variety of hardcore bands identified themselves specifically as 'straight-edge bands'. Straight edge came to mean a lifetime commitment to abstinence, with bands often using their lyrics to promote the advantages of 'clean living'. Beyond the basic tenets of drug-free living, many in the movement embraced vegetarianism, environmentalism, antisexism and antiracism, and other political causes. While still primarily an underground scene, straight edge has spread throughout the world, with vibrant communities from Canada to Argentina, Scandinavia to South Africa and Japan to New Zealand. While many straight edgers eventually 'sell out' and begin drinking, a substantial number maintain a drug-free lifestyle into their thirties and forties.

Like most youth scenes, straight-edge style is not singular or uniform, reflecting the post-subculture theory assertion that 'the relationship between style, musical taste and identity has become progressively weaker and articulated more fluidly' (Bennett and Kahn-Harris 2004: 11). Rather, straight edgers display a wide variety of styles, ranging from the unassuming, preppie emo fashion to the unkempt, bearded, patch-covered style favoured by many crust and gutter punks. The 'youth crew' era of straight edge (1986–91) and its revival (1997–2006) featured an athletic look, including cargo shorts, close-cropped hair, band shirts and hooded sweatshirts—a style that many edge kids over 30 maintain today. The mid-1990s saw the emergence of sports team jerseys, camouflage pants and military caps, while the 2000s brought indie rock-inspired tight pants, bandanas, longer, styled and/or dyed-black hair and black clothes, making some straight edgers distinguishable from their emo/goth/punk peers only by their Xs. Women often wear fitted or sleeveless versions of the same shirts, typically avoiding clothing they see as 'too girly' while still adopting some feminine styles. In the past, straight edgers as a group could have been described as somewhat more clean-cut than their punk peers, and even in 2010 most avoided the more ostentatious markers of punk. However, straight edgers increasingly have followed the ongoing trend of body modification present in music (and other) scenes more generally, displaying multiple tattoos, piercings and stretched ears. Through every era of straight edge and each iteration of its style, the X has remained the movement's most consistent, visible symbol—worn on band shirts, drawn on hands with a marker, stickered on cars, sewn on book bags and tattooed on bodies.

The Strategies of Style

As noted in the Introduction to this book, there is a popular assumption that subculturists grow up and out of their respective scenes, that youthful music communities are a meaningful but relatively inconsequential 'phase' on the path to the more serious endeavours of adulthood. While overstated, such beliefs are not entirely unfounded and might be even more applicable to straight edge, given its strict all-or-nothing ideology: one sip of alcohol forfeits any claim on the identity, and relatively few straight-edge kids remain abstinent past the age of 25. Nevertheless, older straight edgers remain, though they are often less visible than their younger counterparts.

Older straight edgers employ a variety of stylistic strategies, at various times 'covering' or reflexively deploying their edge identity. Some avoid embodying any stylistic references to straight edge despite still claiming the identity, while others proudly sport Xs into their late thirties and beyond, intentionally displaying their edge. Even more straight edgers fall between these extremes, selectively using style in various ways. Bennett (2006: 222–3) notes that 'many of the features attributed by subcultural theorists to young music fans—notably visual style, frequent face-to-face contact, and a publicly articulated collective identity—are not necessarily regarded with the same importance by older followers of rock, punk, and other post-1950s popular music genres.' Individuals are 'reflexive in their appropriation and use of particular musical and stylistic resources' (2006: 223). Not only are the meanings of style more complex than previously theorized, they are contextual and often change over the course of one's subcultural 'career'. The following sections reveal the meanings older straight edgers attribute to their range of stylistic expression, focusing on what they perceive as the limits of style and then on style's potential uses.

The Limits of Style

Nothing to Prove: Straight Edge Gets Personal

In the course of my fieldwork, I observed a trend repeated in a variety of music scenes: younger kids new to straight edge and hardcore spent the most time, effort and resources fashioning a straight-edge style, 'looking the part' of a straight-edge kid. Straight edge was a central identity in their conception of self, an identity prominently displayed on an almost daily basis. In fact, style is part of a scene's initial appeal, as Luke,[1] a 36-year-old tattoo artist, explained:

> At first it was kind of like an image thing. It *looks* cool. It's against everything that high school is about. I got into it that way. Obviously later on I found out that it was more of a positive *lifestyle*.

Ken, now 34, still straight edge and still singing in a hardcore band, voiced similar sentiments:

> Those are the kids that wear straight-edge shirts to every single show and X up at every single show and give other people shit about what they do. That's just like the same thing as skinheads. A fresh-cut skinhead wears his boots and braces and freshly shaven head every single day. Every single day, without fail. It's the same thing. You get into it and you're all gung-ho about it . . . Then after a while it just becomes part of you. In straight edge it becomes less about being the poster boy for straight edge and more about doing it for you.

Most who maintained a straight-edge identity into adulthood shed some aspects of the hardcore 'look', displaying Xs less (if at all) and adopting somewhat more conventional styles even as they continued to adhere to unconventional values. Straight edge became more a personal philosophy than a collective identity or marker of scene status. These older straight edgers identified with the straight-edge *subculture*—that is, a '*cultural phenomenon* that refers to sets of shared values and beliefs, practices, and material objects'—without necessarily participating in the *scene*, the *social* spaces (whether local, translocal or virtual) and relationships related to the production and consumption of (for example) music (Williams 2011: 50). They persisted in their abstinence from drugs and alcohol (and even in some cases enjoyment of the style and music) but often had infrequent contact with straight edgers in social settings such as shows.

Like the punks in Bennett's (2006) and Andes's (1998) studies, older straight edgers claimed to have 'internalized' straight-edge values to such a degree that outward manifestations of their affiliation became unnecessary. Derek, a 37-year-old social worker, claimed: 'I would sometimes X up even if I wasn't going to a show.' However, as he aged, straight edge became more an internalized code and less a style:

> Today [straight edge] is pretty much how I live my life. It is almost like second nature. Whereas before it was something I was very conscious about, I needed to tell everyone I met about it, and let them know about it, I will say in the past sixteen or so years I have mellowed out a lot. It is much more personal to me now.

As a youth, Derek emulated members of straight-edge bands, adopting the styles and Xs featured prominently in band photos on record sleeves. While he continued occasionally to wear hardcore band shirts, and remained very proud of his straight-edge identity, his focus shifted to the *meanings* he associated with straight-edge and hardcore values—creating a personally fulfilling, 'positive', drug-free life in defiance of homogenized mainstream values.

Wearing Xs and Xing up was central to establishing connections and community, but became less necessary (and less possible) as straight edgers aged. Bruce, a 32-year-old geographer, illustrated this shift:

> So, 'back in the day' the community aspect was huge. A lot of that had to do with the collegiate atmosphere I am sure, feeling different from the norm. Now though, I mean even though I still have straight-edge friends near and far, it is much more of a personal thing. Like, when I was younger the team was a huge part. And now I think the personal, internal side is a much bigger piece like, almost to the point where straight edge was something I did, but now it is something I am.

Perhaps because of the rigid behaviour strictures, relatively few straight edgers persist into their thirties, diminishing the opportunity to connect with other similarly aged straight edgers. While all of my participants still had significant relationships with other (older) straight edgers, their scene networks largely had evaporated. If they were to persist, straight edge *had* to become more personal.

Other straight edgers reported that displaying Xs was part of 'proving' authenticity in their youth. As they aged, they saw efforts towards proving their edge affiliation as *inauthentic*; wearing Xs for others' consumption, to gain others' esteem, went against their claim that straight edge was 'personal'. Asked whether he Xed up when he was younger, Tony, a 36-year-old entrepreneur, explained: 'I did in the late '80s, but then I just found that to be a little dumb. I felt the kids that needed to do that need to prove something to themselves, like they needed to talk themselves into straight edge . . . Straight edge was for myself.' Sam, a 32-year-old army staff officer, said: 'I feel that I am old enough and secure enough in life [that] I don't have to impress anyone.'

Kyle, a 32-year-old city planner and Peace Corps veteran, explained why he no longer Xed up his hands, believing his actions better represented his commitment than a subcultural symbol:

> I show [my commitment to straight edge] in my actions. I live my life and all around me know how I live and what I stand for. I have no need or desire to draw on my hands. I loved the power it gave me back in the day, or at least I thought it did. It looked tough and I wanted people to know what I believed in. I didn't have the confidence at that time to just live and show people . . . It isn't really the meaning that changed, it was the need to showcase what I believed in.

Initially, style is central to embodying straight-edge authenticity. As adherents' connection to a scene fades, style is replaced by a more personalized, customized commitment to straight-edge ideology.

Acting their Age: Xs are Impractical

A variety of participants in my original study had altered their appearance for practical—often work-related—reasons. Several working in professional office settings had removed body jewellery and let piercings close over. Even those who still strongly identified with straight edge found the style occasionally inconvenient. For John, a 33-year-old father who still played in a hardcore band, his work context made Xing up at shows difficult, even if he might want to do it:

> I work in a professional environment where walking around with Xs half faded on my hands looks a little weird. That's mainly it. I have a Sharpie[2] in my guitar case, and one in my car, but most of the time, it comes down to 'I have an 8.00 a.m. meeting tomorrow with someone, and I don't want my hands to look like I went to some bar the night before.'

Maggie, a 31-year-old public librarian, agreed. While she had consistently Xed up in her youth, her work environment was not conducive to such displays:

> I suppose when I was getting older and didn't want to go to work and have to explain to a 50-year-old why I had huge, faded Xs on my hand . . . There's also a confidence that comes with age. You don't really have to put yourself on display anymore. You're just comfortable with who you are.

Similarly, some straight edgers simply tired of explaining their abstinence. Younger and embedded in a scene, their choices needed less explanation. Surrounded by un- or ill-informed workmates who assumed everyone drank socially, older straight edgers found it more convenient to avoid straight-edge symbols, thereby avoiding long explanations. Bruce said:

> I pretty much never claim edge. [If someone offers me a drink] my answer is always, 'I'm not really into it' because either someone has heard of straight edge, and it is in a negative way or SO out of whack, or they have no idea. And frankly, after so long, I am so tired of holding people's hands and explaining that it is OK not to drink or that there are—[GASP]—lots of people that don't.

Other straight edgers did not necessarily judge such strategies negatively, as indicative of reduced commitment. While public displays or declarations of straight-edge affiliation might be appreciated, evaluations of commitment rested more upon adherence to straight-edge values and practices.

A few older straight edgers worried about being perceived as trying to look younger than they were—that younger adherents might see them as 'trying too hard' to fit in. Asked if he ever worried about being perceived as trying to look too young or too 'hip', Tony said: 'Yes. Always. I never wanted to be that 40-year-old man in skate gear.' His personal style, he explained, consisted primarily of underground streetwear designers mixed with more conventional brands. This more muted style allowed him to feel 'different' without appearing to be 'stuck' in a style that seemed too 'young' for his ageing appearance.

Circumstances made some straight edgers 'act their age' while others toned down their style to avoid the (real or imagined) judgement of younger fans. Even those still passionate about straight-edge and hardcore music considered how an ageing body might one day lead them to feel too old to look the part of a 'scenester'.

More than Music: Critiquing the Edge

The third general explanation participants gave for muting or abandoning their straight-edge style was that style somehow undermined their focus on living straight-edge values. For most older participants, straight-edge identity constituted more than abstinence. The spectacular aspects of the straight-edge music scene detracted from one's commitment to the straight-edge values of individuality (Williams and Copes 2005). Fashion stole attention from values and practices associated with straight edge: vegetarianism, individuality and positivity. In short, participants claimed that *image* undermined action, as straight edge became more about what adherents did rather than what they wore. Tony said:

> I really don't identify with straight-edge kids any more for I am no longer a part of the straight-edge music scene and I feel that straight edge puts too much on the music. So I find myself feeling like an outsider in a scene that drives itself on the image of the

young . . . I feel like we need to embark on a new chapter as we get older and that's what I am trying to do. Move straight edge for myself as a lifestyle not a *music* movement.

Derek, who enjoyed wearing shirts of 1980s era hardcore bands, acknowledged the limitations of style and saw straight edge as having potential beyond a music scene:

[Straight edge] was just a building block to something more. We thought that we had an edge on the drugged out hippies, that we could really use straight edge to do something more like combine crust punk politics with straight edge.

Touring in bands during the 'political' era of straight edge in the early 1990s, Alec, a 39-year-old college instructor, had witnessed how straight edge played a role in kids' political development, and how the scene eventually changed: '[People realized] there's these bigger issues. There were a lot of people that made incredible political decisions based upon their involvement in the scene. But by '97, '98 . . . it was all about being *straight edge*.'

Alec and Kyle considered labels articulated via style symbols to be superficial compared with *action*. Alec said: 'A label is nothing without practice in the community,' and Kyle remarked: 'I have so many different facets that it plays a role in my life but it's on the back-burner. I don't define myself as edge anymore, although I am. I live the way I think I should, I live the way I want others to live.' Older straight edgers especially did not want to be perceived as being into straight edge for the 'image'. Jason, a 37-year-old video conferencing engineer who was virtually indistinguishable from any other corporate office worker (so long as his many tattoos were covered), believed transcending style was an important step towards maintaining a commitment to the identity in the long term:

You can't really put anything on, you can't really sell a straight edge image, you can't sell the lifestyle. You can sell it on a T-shirt, you can write 'straight edge stormtrooper' on it and sell it at Hot Topic, but that doesn't mean this person is really going to be around. Everybody I hung out with in my life, back home when I was growing up, the kids who got me into hardcore, *none* of them are straight edge anymore. There's not a single one of them. I'm still friends with them, but none of them are straight edge. These are kids who would X up at every show.

Finally, while these straight edgers were disappointed by what straight edge had failed to become, others were embarrassed by what they perceived straight edge had in their view *actually* become: a judgemental, male-dominated, apolitical, hypermasculine club. They purposively *avoided* the markings of straight edge to avoid being identified with such characteristics. Indeed, the more 'militant' strain of straight edge plays a role in many adherents' eventual abandonment of the identity (Torkelson 2010). While critics have decried the scene's hypermasculine tendencies since its inception, many point to the mid-1990s as the beginning of the scene's most militant era—also the era in which my participants graduated into adulthood. Tony asserted that he avoided publicly displaying straight edge due to the negativity associated with straight-edge 'tough guys', such as those in the hardcore crew FSU: 'The FSU thing gives it a bad name and I have worked too long and too hard to come up from nothing to let some dumb ass angry white kids from Boston make me look bad.' Likewise, Alec said: 'Later in life [straight edge] became virtually almost like an embarrassment. When you're watching late night TV and there's a straight edge gang on there, you're just like, Oh man. What happened?'

Shifting from scene participation, most older straight edgers described their straight-edge identity more as a personal lifestyle, reflecting 'individuality, self-expression and stylistic self-consciousness', and guiding their everyday practices, tastes, consumption habits and leisure activities (Featherstone 1987: 55). Many shifted focus from crafting a spectacular image to enacting their subcultural values in 'adult' contexts. Their continued connection to or dissociation from straight edge reflected their experiences coming of age within a particular moment in straight-edge history. As such, their interpretations of style and their strategic avoidance or deployment of straight-edge symbols were temporally positioned, with their present embodiment of straight edge echoing memories of the past.

The Potential of Style

Style as Community: The 'Old Guy' Club

While most older straight edgers tone down showing their straight-edge affiliation, only a few avoid displaying straight-edge identity altogether; the majority strategically use straight-edge identifiers on occasion to connect with a meaningful past, communicate longevity and/or authenticity, set an example for younger kids or symbolically resist social norms. Indeed, the same individual straight edger may hold seemingly contradictory views regarding style displays, seeing style as unnecessary or irrelevant while also periodically finding satisfaction in representing straight edge. For some older straight edgers, wearing a shirt from a show that took place ten years ago indicates an overemphasis on image; for others, wearing such a shirt symbolizes a connection with an important time in their life or resistance to perceived mainstream culture.

The first explanation participants gave for continuing to show their straight-edge identity was that embodying straight edge reminded them of a meaningful time in their youth and connected them with a community of similarly aged straight-edge peers. Wearing an old band shirt felt 'comfortable' and evoked a sense of identity, even if the older straight edger involved interacted with straight-edge contemporaries primarily online, where presumably choice of clothes should not matter. Style helped facilitate an 'imagined' community for those straight edgers who rarely attended shows or encountered other straight-edge kids. Derek said:

> I feel like there is sort of a community of older hardcore dudes [online]—it's almost like a therapy or support group for people from the late '80s who like Warzone and Youth of Today . . . Most of us are the same—we have straight jobs and wear suits and stuff to work, but we break out the shorts and Vans on the weekends . . . My wife says that everyone my age that was into hardcore or straight edge at the same time still dresses the same, it is almost like the 'old guy' uniform. I would feel silly if I tried to get into the clothes and music kids are into in the scene today.

Several participants wore band shirts, but primarily shirts from 'their' era of hardcore—an era perceived as more authentic than hardcore's recent, often metal-influenced sound. Clothing choices became strategies of authenticity, but also represented nostalgia for an intensely meaningful time in their lives. Derek continued:

> I mean it is weird, as I have gotten super old, I have drifted back to wearing the same clothes I did when I was sixteen. [I wear hardcore shirts] all the time. But the bands I wear are all from my era—or before . . . I think that is what I feel the most comfortable in. I wear DYS shirts because I like them, and I was into them as a kid—I wear Unit Pride shirts because I saw them, I played a show with Judge, etc. I feel a personal connection to that time period.

John still collected old T-shirts, buying them on eBay and wearing them not only on weekends but also under his work clothes. He also saw style as a way to resist corporate fashion:

> I wear hardcore shirts under my work shirts. I don't see the reasoning behind wearing a shirt that says 'Abercrombie' across the front of it. You're a walking billboard for some giant corporation. I would rather support a band that I recently saw, or a band whose music and/or lyrics I find inspiring.

Demonstrating the fluid meanings of style, some straight edgers avoided straight-edge markings as they critiqued straight edge in their mid-twenties, only to return to the style as they aged. Sam went through a period when he rarely displayed his straight-edge identity. However, when he deployed, straight edge—and straight-edge style—took on renewed significance:

> I Xed up a lot through high school and up towards the end of college . . . and then the edge scene sort of died and it was just goofy to X up. So for years no Xing up. So go back a couple of years and I start going to Iraq and Afghanistan . . . I started wearing more edge shirts around base and putting edge type stuff on my kit. I needed to hold on to something that was near and dear to my heart . . . something at my core . . . I had my

> wife's ring around my neck with my dog tags and I wore XXX shirts when I wasn't out on a mission. And over the last couple of years I have started to flaunt the edge with shirts and stuff more and more. I am 32, and I don't care . . . that's who I am and I have started to take more and more pride in it.

Even though very few of these straight edgers regularly participated in a physical music scene, their straight-edge peer group and past history still served as stylistic reference points in certain contexts, symbolizing their internalized affinity to the identity.

Style as Legacy: Flying the Flag

Several straight edgers believed that wearing Xs or other straight-edge identifiers helped set an example for younger adherents or helped keep the movement's legacy alive. Demonstrating that some straight-edge kids maintain their commitment as they age—rather than being 'true til 21'—was important to them. Luke, who had Xs tattooed on his hands and 'Lifer' across his throat, said:

> [Straight edge] kind of blossomed into, you've *got* to set an example, you've *got* to show kids what it's about. 'Cos if *you* don't, then who's going to? The strong ones who are going to last need to set the example or else it will just be prostituted out to whatever fits their lifestyle.

Given that relatively few straight edgers maintain the identity into their thirties, those that remained sometimes felt a responsibility to demonstrate their longevity. Derek had few older, positive straight-edge role models as a kid: 'I just try and be a good example that it is possible to get older and not sell out. That example wasn't around for me as a kid—and I would have liked to have someone older to look up to who hadn't totally given up on what they believed in.' Derek's ageing body, combined with straight-edge styles, became an asset to living out his straight-edge values.

Explaining how he wore shirts with straight-edge messages such as 'It's OK not to drink' in public spaces, John conveyed that having to explain his choices might help people question peer pressure:

> It's weird trying to explain to your 40-year-old co-worker who grew up listening to INXS why you don't drink. It's easier to just go incognito I guess. [But] I could give two shits what they think. Plus, maybe I empowered some younger kid to not succumb to peer pressure with regards to drinking. I wore the shirt to a bachelor party, and I had a bunch of people come up to me to say 'I used to be straight edge' or 'I can't believe you're still straight edge' like it was weird for someone to be in their thirties and still be edge.

While he often dressed relatively conservatively, Jason also believed it was his duty to share straight edge when he went to shows:

> I truly believe in the message and ideas of straight edge and making sure that the tradition and movement lives on. It's very powerful. Think about how many kids out there feel lost in junior high and high school. The boys and girls looking for something to believe in, something to make them feel less alone.

Although he did not feel especially compelled to set an example for younger youth, Sam specifically Xed up for reunion shows featuring old bands, showing his peers that some straight edgers maintain their commitment into adulthood:

> So now [straight edge is] something I do for myself . . . but I always do it for my friends. I feel I have a responsibility to keep it alive . . . like we are a dying tribe and [we] keep the edge alive . . . But I am more looking to show who I am to my age/peer group versus the younger ones. Like going to see the Earth Crisis reunion. It was like, 'Yup, I am a kid from back in the day and I am still edge.'

Some older straight edgers strategically used their ageing bodies as markers of authenticity, transforming the meaning of ageing from being a detriment into an asset.

Style as Resistance: Down for Life

Finally, a minority of older straight-edge kids continued to boldly display their straight-edge identity, *consistently* wearing shirts with Xs, straight-edge slogans and hardcore bands, or permanently altering their bodies. They embodied a symbolic, stylistic resistance to 'adult' life, even while undertaking many adult roles and responsibilities. Given that their style displays increasingly occurred in adult settings, the potential stigma—and opportunity for meaningful resistance—may have been greater than in their youth. Several reported that they felt even greater pressure to drink as adults in professional settings. Younger straight edgers may claim to alter their bodies in defiance of adult culture, but upon reaching adulthood most acknowledge that they directed their choices primarily to the scene. For older straight edgers, no longer as embedded in a physical scene, stylistic resistance took on new meanings.

Some straight edgers cultivated a relatively conventional appearance, adding pieces of subcultural 'flair'—a belt buckle, hipster haircut, vintage sneakers, slightly stretched ears, black clothes or a small tattoo. Such affectations were personally meaningful and noticeable to others, but quiet enough so as not to disrupt or damage adult legitimacy. John, for example, wore his 'X' Swatch-brand watch, popular amongst straight-edge kids since the late '80s, a marker instantly recognizable to anyone familiar with straight edge but simply a quirky or odd item to outsiders. Sam explained how he enjoyed the 'little acts of resistance', saying, 'The demon I was resisting in high school was HUGE; now it's not so huge . . . but I still like to resist.' Bruce discussed how living and working in very conventional settings made small acts of embodied resistance even more important:

> If being straight edge or vegan or any kind of counterculture teaches you anything it is that, oh my god, sit down, you don't have to live your life one way and that there are a lot of rules, or not even rules but perceived rules that people freak the fuck out over for no reason . . . I have no problem working on a military base or in an office for that matter but it doesn't mean I am going to wear khakis. Why the fuck would you wear that shit? SO BORING!

Still other straight edgers exhibited less easily hidden markers of straight edge. The growth of the tattoo and body modification subcultures from the 1990s to the present has meant that thousands of straight edgers have Xs adorning their bodies. Many straight edgers can cover their tattoos for work and certain social functions. However, a few have Xs tattooed on their hands, neck, fingers or even face, symbolizing pride and the permanence of their commitment. Other examples of tattoos include 'drug free', 'stay true' and 'str8 edge' across their knuckles. Others have significantly stretched ears that, while not exclusively associated with straight edge, still mark the bearer as an outsider. All but two of my participants had tattoos, and several had full sleeves on at least one arm or tattoos on forearms and other prominently visible places. Brian, a 32-year-old graduate student finishing his PhD in biology, explained the importance of using style to stand out from the crowd:

> I've always disliked mainstream fashion. I would try like hell to set myself apart, to be different in my own way and the clothes I would wear is how I expressed myself. Now that I'm in academia, I still wear my old band T-shirts precisely because no one knows who they are and I love that. That person has no idea who I am and their judgment of me is most likely way off base. I revel in the fact that when they get to know me they are so surprised that I am where I am. That is probably one reason why I have a [tattoo] sleeve.

Though he refused to X up or wear band shirts to hardcore shows, Brian repeatedly emphasized how he used style and tattoos in *other* contexts to distance himself from anyone who would judge him on superficial characteristics:

> It's also really important for me to be completely different to the Radiohead/311 listening people of the world. If I apply for a job and they just judge me for having a sleeve and make a decision based on that judgment, then fuck them, they don't get to have me as a colleague.

Maggie had a full sleeve of tattoos, including XXX with 'pure' written on her wrist and 'Promises Never Broken' across her back. Asked if it was important for her to set herself apart from 'everyday' people, she replied: 'Yes. It still is. I look at so many people around me and find them to be incredibly boring. Is that mean? Probably, and I don't really know what most people are like. But they almost seem like they're just existing instead of living.'

Even while questioning an overemphasis on style and its connections to scene politics, older straight edgers saw potential in carefully embodying straight-edge symbols. Style helped them construct an imagined community, often based upon past experiences but oriented towards present and future straight edgers. Rather than being a reminder of a failure to 'grow up', an ageing body can be an asset, strategically displayed within specific contexts towards particular ends. Embodying the *subcultural* style markers of straight edge was occasionally important, even though participation in a social *scene* was not.

Conclusions

This chapter has illustrated the great variety and complexity of the meanings older subculturists construct around style. Individuals hold multiple, sometimes contradictory interpretations of straight edge that are reflected in stylistic choices. Straight edgers over 30 carefully (re)considered the meaning of straight-edge symbols in their lives. Rather than abandoning spectacular style altogether or constantly displaying their edge identity (as many had in their youth), most straight edgers reflexively and strategically represented straight edge to suit the demands of a given context. Style was a symbolic marker of deeper feelings about the straight-edge identity. Rather than a mark of failure to 'move on' to adult pursuits, representing straight edge gave many adherents a sense of pride that they had maintained their commitment while so many others had not. The participants in the study had all created lives beyond the music scene, careful to avoid becoming 'stagnant punks', inappropriately hanging on to the trappings of youth (Davis 2006). Noting the limits of subcultural style and of straight edge more generally, most redefined straight edge in more personal terms, and most saw potential in style beyond achieving status in a scene.

While *all* older straight edgers expressed that straight edge had become a more personal philosophy, they differed in how best to put this insight into practice. For some, toning down or even abandoning straight-edge fashion was central to their conception of straight edge as an internalized philosophy, a stage Andes (1998) calls 'transcendence'. For others, authenticity required displaying straight-edge symbols, not necessarily for collective consumption in a music scene but to visually separate them from mainstream culture, continue to challenge drug and alcohol culture, and honour the legacy they cherished.

Embodying straight edge neither necessarily upholds nor negates one's sense of authenticity. Rather, authenticity is an ongoing project with a contextual meaning. For older straight edgers especially, legitimacy hinged on *action*, not image—primarily remaining strictly abstinent but for many also pursuing vegetarianism or other forms of lifestyle politics. Spectacular style is often a mark of status for subcultural youth, but for older straight edgers style was not *essential*, even when it was meaningful. Overemphasizing the style component of music subcultures may inadvertently exclude the study of older adherents for whom style is no longer central to their subcultural self. As Clark (2003: 234) suggests about ageing punks, labels eventually become limiting and action replaces fashion: 'The threatening pose has been replaced with the actual threat.' Older straight edgers re-emphasized individuality while honouring their roots, an individuality *professed* in their youth but *lived* in adulthood.

–2–

Performances of Post-Youth Sexual Identities in Queer Scenes

Jodie Taylor

Introduction

The physicality of an ageing body is inescapable, and for this reason the processes of ageing and the signs that mark ageing identities often appear as something beyond one's social and cultural self, something innate. While it has become commonplace—at least in a post-structural context—to argue that one performatively assembles selfhood, age appears bound to the temporal norms that intersect with the corporeal signs of passing time. However, because ageing is a process embedded in the social and cultural, discourses of ageing construct normative and healthy modalities of ageing while rendering the nonnormative unhealthy, undesirable or deviant. Much like the physicality of the human body, which through a cultural reading is sexed—in turn 'determining' a normative and thus socially appropriate performance of gender (Butler 1993)—the physical signs of an aged body are also interpreted through a cultural lens. Age, like sex, corresponds to a set of social and cultural norms that sketch out desirable, culturally intelligible and thus successful performances of how one should mature. Just as a female body becomes normalized through its reiteration of femininity, an ageing body similarly becomes normalized through the reiteration of prescriptive age-appropriate behaviours.

For example, in a white, Western, heterosexual, working-/middle-class context, it is deemed age-appropriate for youth to invest heavily in popular music, scenic identificatory practices and displays of concomitant style. It follows that, as one moves beyond the category of youth, popular music scene participation should become less potent, giving way to the normative foci of a post-youth life course—careers, marriage, reproduction and the like. As one commits to a career, marries or reproduces, there is an expectation that one will also start to look and act more like a normative adult. The hegemonic image of a successful adult necessitates doing away with the more obvious displays of youth that might be evidenced by dressing in alternative or extreme forms of clothing, wearing radical haircuts, retaining facial piercings, attending all-night dance parties or following new music. Yet, for some people, youthful forms of scene participation and a commitment to the visual and musical styles associated with a particular scene may not necessarily diminish with age. Those who continue their scenic investments can thus be seen as troubling this responsible progression towards maturity by favouring a prolonged youthfulness and a lingering within early adulthood.

Queer scenes can offer one such example of 'nonnormative' ageing, where typically youthful and less coherent social activities such as attending gigs and dance parties, nonmonogamous sexual play and recreational drug-taking may continue to remain prevalent and more central to one's lifestyle choices than marriage, reproduction and child rearing (Halberstam 2005; Taylor 2010). Of course, not all queers resist the reiteration of prescriptive age-appropriate behaviours, but some do, choosing to depart from the normative temporal frames of marriage and kinship family, and create alternative modalities of post-youth scenic identification. It is at the intersections of ageing, sexual identity and scenic participation that I take up my argument of queer ageing.

In this chapter, I am concerned with the ways in which post-youth queer subjects construct and perform age in the context of scene participation. To begin, I will survey the current literatures on ageing within music scene contexts, elucidating the problematic relationship between popular music and the post-youth subject. Employing a queer theoretical framework, I will then unpack the idea of heteronormative temporalities and investigate the more recent emergence of homonormativities that place a different, yet equally limiting, set of expectations upon the ageing sexual subject. Positing the notion that one may 'queer' performances of ageing—resisting both hetero- and homonormativities—I explore performances of ageing queer identities in relation to ethnographic research on queer scene participation in Brisbane, Australia, suggesting that queers problematize normative constructs of middle age and age-appropriate behaviour. My empirical case study will focus on these 'improper' presentations of middle age in much the same way that previous queer theoretical work has addressed subversive gender and sexual performances. The chapter will then conclude with a discussion on how age may also be understood as performatively constituted, and how scene participation figures within this.

Scenes and the Ageing Subject

Popular music is often perceived as dependent upon communities of youth to propel and invigorate fan bases and style, while the aesthetic value of music for older fans is commonly passed over as nostalgia, and at worst aligned with discourses of failure whereby subjects are understood to have delayed or failed to progress into 'proper' adulthood (e.g. see Thornton 1995; Weinstein 2000). This youth-centric master narrative has obscured the potential for, and evidence of, popular music's significance as a component in the performative assemblage of post-youth forms of identification. In part, this has to do with the centrality of youth to stylistic forms of cultural resistance perpetuated by an approach to subcultural theory taken by the Birmingham Centre for Contemporary Cultural Studies (CCCS).[1] Since the limitations of this approach are now well-rehearsed debates (Bennett and Kahn-Harris 2004; Hodkinson 2002; Muggleton and Weinzierl 2003), I shall not reiterate them here. Rather, I have chosen to employ the theoretical concept of 'scenes' (Bennett and Peterson 2004; Straw 1991). I find the concept of the scene particularly useful, as it enables us to look beyond the more spectacular leisure pursuits of youth and incorporate new vernacular knowledges that signal the types of identification, scenic competencies and cultural politics that evolve from the prolonged interactions of post-youth subjects who may participate within and across local, translocal and virtual contexts.

With limited exceptions, ageing subjectivities and music scene participation constitute a poorly theorized phenomenon, and empirical studies on this topic are quite scarce. For example, Andes (1998) suggests that just as one becomes punk, it is also an identity that one must shed as one ages, for it is only those with creative or organizational authority within the scene who are likely to maintain—or indeed capable of maintaining—involvement. Reasons for this go unexplored by Andes, however. In Davis's (2006) study of punk scenes, she identifies four typologies of ageing punks, offering what she terms 'successful' and 'unsuccessful' examples of adulthood and scene identity synthesis. While Davis's study offers some relief from the narrative of failure that we find in earlier studies, such as Weinstein's (2000) work on ageing heavy-metal fans, failure remains imbricated in a dialectic that rationalizes participation as either successfully or unsuccessfully integrated into a version of normative adulthood. Meanwhile, Gregory's (2009) discourse analysis of ageing female rave participants from Toronto demonstrates how, for the majority of her sample, post-youth rave participation conflicts with normative perceptions of adult responsibility such as committed partnering, motherhood and ageing corporeality. Through discursive analysis, Gregory concludes that '*age*—like *gender*—is a discursive concept, often used to marginalize certain activities and/or identities as immoral and/or "deviant" ' (2009: 73, emphasis in original). This, she says, is implicit in the way her respondents talk about their withdrawal from the scene.

The aforementioned studies suggest that, for the most part, scenic withdrawal is a by-product of the ageing process. However, as a handful of scholars have shown, adulthood and music scene participation are not necessarily mutually exclusive. In Kotarba's (2002) study of ageing fans of rock 'n' roll—a music responsible for significant post-war cultural battles between adolescence and adulthood—he argues that in the same way that rock 'n' roll inspired youth identities, it continues to give critical meaning to adult lives. As Baby Boomers age, have children and then grandchildren, and develop new social relations, rock 'n' roll facilitates the incorporation of the values, tastes and sensitivities of their youth into the ever more complex everyday subjectivities of adulthood. In Vroomen's (2004) study of ageing Kate Bush fans in a virtual scene context, she argues that 'varying degrees of "sceneness" ' (250) persist in the lives of the less spectacular ageing female fans, whose commitment to Bush becomes more personalized and confined to private moments and spaces. In Smith's (2009) account of an ageing British Northern Soul scene, she explores returning and continuing participation in scene activities by ageing individuals, and looks at the age-derived conflicts that ensue. The appeal of scene participation, she argues, can be considered 'the same in adulthood as it was in youth—to possess an identity and a form of cultural involvement that results in the achievement of scene-specific status, personhood and subsequent selfhood' (436). Finally, in Bennett's (2006) study of ageing punk fans from East Kent, England, he demonstrates the multiple ways in which ageing fans continue to legitimate their status as punk through modifying their aesthetic and discursive practices so that ageing may be modulated in ways that these participants find advantageous and more conducive to adulthood.

With the exception of Vroomen's (2004) and Gregory's (2009) work on heterosexual female scene participants, the studies discussed above predominantly deal with post-youth scene involvement by heterosexual male participants. As such, the case examples that I will offer later in this chapter add a significantly new dimension to the literatures on music and ageing through a discussion of the implications of ageing and continuing scenic commitment for both males and females who sexually identify as queer.

Normativities of Ageing

Before exploring the everyday and lived experiences of post-youth queer subjects, and the multiple ways in which they subvert many of the social and cultural normativities associated with ageing through scene participation, it is necessary to first explicate the discourses that interpellate ageing subjects. When looking to social gerontology, lesbian/gay/bisexual/transgender/queer (LGBTQ) studies and cognate fields of research, certain standards pertaining to respectability, social expectation and age-appropriateness are inflicted upon ageing subjects. Often when someone falls short of meeting these standards, his or her intelligibility as a culturally coherent subject is weakened, blurred, queered. 'Queer,' argues Warner (1993), 'gets its critical edge by defining itself against the normal rather than the heterosexual' (xxvi). Following Foucault (1976), who stressed the revisionist potential of discursive strategies, queer is a deconstructive project that moves beyond minoritizing logic and assimilatory discourses of LGBTQ tolerance, normalization and legitimization (Sullivan 2003). Queer offers a theoretical perspective from which to challenge the normative, thus all normalizing regimes come under attack within this framework (Warner 1993). Just as queer has proven highly useful in the critique of gender and sexual normativity, it also lends itself to critiques of the temporal normativities that, among other things, produce successful and subsequently deviant ageing identities (Dinshaw et al. 2007; Halberstam 2005; Sandberg 2008).

Heteronormativity, a concept beholden to queer theory, is the set of norms that legitimize and privilege heterosexuality as natural, coherent and moral through situating homosexuality in negative binary opposition to it (Sullivan 2003; Warner 1993). Heteronormativity structures our understanding of the default alignment of biological sex, gender identity, gender roles and sexual desire (Butler 1993). Moreover, it relates to a 'sex hierarchy' in which certain approaches to sexual activity—such as marriage, monogamy, heterosexuality, reproduction and sex in the privacy of one's home—are considered normal and therefore good. Meanwhile, other practices—for example, homosexual, queer, interracial or intergenerational sex, promiscuity, BDSM[2] play and group sex—are considered abnormal and therefore bad (Rubin 1984).

'Age as a social category consequently plays an important role in how heteronormativity functions,' argues Sandberg (2008: 131). In many ways, what constitutes 'normal' and thus 'successful' ageing is tied to notions of heteronormative temporality, which demands mandatory continuity (Halberstam 2005). This is where the age-linked transitions of a successful life-course can be mapped in terms of a linear, sequential progression from birth, childhood and adolescence through early adulthood, career development, marriage, reproduction, child-rearing, retirement, old age, death and kinship inheritance. Social institutions—taxation, social security benefits, marriage and kinship family rights—usually reward those who stay within this timeline. One must not only work hard and reproduce, but also aspire to reach old age; thus it follows that activities perceived to inhibit one from reaching a 'healthy' old age, such as sexual promiscuity and recreational drug-taking, are also condemned—especially when one continues to engage in these activities beyond the excusably irresponsible time of youth (Halberstam 2005; Sandberg 2008).

This heteronormative temporal map is highly incongruent with the lived, nonmatrimonial, childless realities of many queer post-youth subjects, who invest considerable time in scene activities such as playing in bands, attending gigs, DJing, dancing in nightclubs and running events, seeking multiple sexual partners and taking recreational drugs well into their thirties, forties, fifties and beyond. According to Halberstam (2005), the 'stretched-out adolescence' of 'queer culture makers' disrupts 'conventional accounts of subculture, youth culture, adulthood, and maturity' (153). When the break between youth and adulthood—marked by marriage and reproductive time—lacks clarity, it can cast significant doubt over the security and intelligibility of heteronormativity. Thus, whether enacted by a homo- or heterosexual subject, the stretched-out adolescence that characterizes forms of post-youth scene involvement is potentially as troubling to heteronormativity as the queer body itself.

Of course, not all queer people wish to, or are able to, invest in scene activities in the years following youth. Where law and finances permit, some choose to enter into legalized same-sex unions and, with the aid of reproductive technologies and/or adoption, may also choose to have replica hetero families, subscribing as well as they can to the dominant social organization of life time. For some, such as those who live in rural or regional areas, desirable scenes—scenes in which they are free to express their sexuality in a manner of their choosing—do not exist, so scenic participation has never been an option. For still others, the thought of accessing a scene may inspire anxiety, fear and feelings of inadequacy. An increasingly common reason for the latter is often a response to the homonormativities of the gay public sphere (Duggan 2002). In a roundtable discussion on queer time published in *GLQ* in 2007, Hoang (cited in Dinshaw et al. 2007) reminds us that 'there is also a homonormative time line. We pity those who come out late in life, do not find a long-term partner before they lose their looks, or continue to hit the bars when they are the bartender's father's age. We create our own temporal normativity outside the heteronormative family' (183–4). Homonormativity intersects with heterosexism to produce negative impressions of lesbian and gay ageing, whereby the subject becomes less desirable as he or she ages and the lifestyle associated with their sexuality—hitting the bars and cruising for sex—is seen as unsustainable in mid and later life (Casey 2007).

The current literatures on LGBTQ ageing present overwhelmingly gloomy, socially awkward and disadvantaged accounts of middle and old age—accounts that deal mostly with the male perspective (Bergling 2004; Cruz 2003; Herdt and de Vries 2003). While these studies have helped to combat the heterosexist bias in the field of social gerontology more generally—as Plummer (2005) points out—at times they also incidentally reinforce the tired gay joke that anyone over the age of 30 is old. There are many pejorative stereotypes of older lesbians and gays, such as the 'tired old queens', 'desperate old poofs', 'dirty old men' or 'bitter old dykes' who are seen to 'prey' on the young (Kiley 1995; Price 2007; Shively 1980). Or they are the old, sexless rejects who have been shunned by a gay mainstream obsessed with youthful, attractive bodies. Price (2007) similarly reiterates that the assumed position of older gays and lesbians outside of a kinship family structure means they are depicted as shamed, resentful and banished by 'their own kind' to a life of loneliness—a further warning of the perils of perversion. Informed by personal experience, Shively (1980) recounts how middle-aged gay sexuality is considered repulsive to the young: as Shively (1980) argues, 'Aging in a gay community generally does not reassure you that you are still worthwhile, loved, wanted or a human being' (75). While it is both unfair and an overstatement to broadly charge young scene participants with thinking in this way, unfortunately we continue to see this logic further underscored in the mainstream international gay media. In an online feature article from *Out* on Boy George's and George Michael's recent drug-fuelled sexual exploits, the opening line to the story reads: 'What is it about middle-aged queer British pop stars from the 80s? Why can't they settle down, keep their noses clean, their peckers zipped, and their faces out of the papers?' (Simpson 2009). It is not the drug use or sexual activity that offends in this and similar commentary that one finds in mainstream gay press; rather, in most cases, it is the age of the doers.

Magazines and other media that feed into pink dollar[3] economies increasingly perpetuate these homonormative images and identities, which we see reflected in the gentrification of gay public space. As a result, some scholars argue that the Western mainstream lesbian and gay scenes that occupy urban commercial spaces are becoming more exclusionary and intolerant of that which doesn't conform to normative ideas of youth and beauty (Binnie 2004; Casey 2007; Taylor 2008). Of course, this is not solely a LGBTQ problem, as the value of the ageing body is restricted in contemporary Western cultures generally. However, as I have argued, there is evidence to suggest that ageing people who engage with mainstream urban gay scenes are particularly at risk of age-based exclusion. Whittle (cited in Casey 2007) reflects this, noting that, 'ironically, beautiful young people don't need safe and tolerant places—because sex is always going to be easy for them; they are, after all, beautiful and desired' (131). What were once constructed as spaces of 'refuge and tolerance' are, according to Taylor (2008), 'increasingly theorized as "niche" markets for development and exploitation . . . The pleasures and problems of scene spaces are sharply highlighted in the idea of "buying into" commercialized leisure spaces . . . [and] ultimately "selling out" politicized identity credentials for marked-based purchasable ones' (524). For Binnie (2004), Casey (2007) and Taylor (2008), this process determines who, on the basis of race, ethnicity, gender, class and age, is included and who becomes the 'queer unwanted'. Disqualified from heteronormative and homonormative status, what happens to the post-youth, sexually and socially adventurous, unmarried and 'unhip' scenester? An emphasis on middle-class consumption, youth and the limitations of style that mark many commercial lesbian and gay scenes renders these spaces undesirable and sometimes inaccessible to the ageing queer scene participants I interviewed for this research, with many of them—such as Tex—telling me: 'I wouldn't feel comfortable at the mainstream gay clubs . . . plus I like good dance music and yeah I'm often disappointed by "gay" clubs with the music.' However, as I will show, ageing queers such as Tex continue to make and vibrantly participate in certain queer scenic formations outside of what some of those within alternative queer scenes might consider the more mainstream gay spaces. Undertaking numerous complex identity negotiations, queer scene members produce their own ageing narratives and stylistic modalities to affect a decidedly queer use of lifetime.

Queer Scenes: Ageing Queerly

As the aforementioned literatures suggest, ageing scene members periodically construct particular kinds of musically embellished ageing identities. Furthermore, queer scenes allow for a redefinition of the adolescence/adulthood binary, 'precisely because many queers refuse and resist the heteronormative imperative of home and family . . . prolong[ing] the periods of their life devoted to subcultural participation' (Halberstam 2005: 161). Turning now to a queer club scene in Brisbane, Australia, I will explore the cultural dynamics, musical and extramusical style and discourses of ageing produced in this scene. Much like Hodkinson (2002), in his study of goth, I am a 'critical insider researcher' who has been a part of queer scene cultures for a number of years predating the study (Taylor 2011). Research on this scene, which was carried out between November 2005 and April 2009, has involved lengthy semistructured interviews (recorded and transcribed) with nineteen queer-identified people. This is complemented by countless hours of scene participation conducted at various parties, nightclubs and gigs in Brisbane and further augmented by fieldwork undertaken in translocal queer scenes in Sydney, Berlin and London. All personal information included here is with the respondents' consent, and pseudonyms are used to ensure respondents' anonymity.

In this account, I will make particular use of data synthesized from ten queer-identified middle-aged interviewees (aged between 38 and 46) who all exhibit dedicated post-youth scene involvement and assume a variety of scene-related roles including participant, organizer, DJ, musician and cabaret performer. Five interviewees are female and the other five are male. Nine of these are Caucasian and one is Indigenous Australian. Eight are clearly identifiable as middle class and two are working class. None of the interviewees was married or had children. All had full-time (or equivalent) employment across a variety of fields, including health work, academia, business, administration, information technology and the arts. These socioeconomic indicators can be considered representative of the scene at large, which is made up predominantly of a similar number of male and female,[4] middle-class Anglo Australians, who can be located across various employment sectors.

While, for the uninitiated, the term 'queer scene' may not appear to be intrinsically tied to a set of musical aesthetics, in the same way that the notion of a goth scene or hip hop scene is, music and dancing are paramount to queer scene gatherings. According to local scene participant Peter:

> Music has been the binder of the queer community over the years, because the queer scene has been just that, a dance scene, a music scene. When people get together and go out for a night it's music that they go to . . . Music binds social outings especially for the queer community because the scene was burgeoning in the clubs where you could escape from the rest of the world.

Like other queer scenes globally, Brisbane's scene congregates around music, with performing, appreciating and dancing to music being the most common forms of collective activity in which scene members engage. While electronic dance music (EDM) is the style most typical of scene gatherings, it is not uncommon within the Brisbane scene, and in other scenes that I have observed in London and Berlin, for live bands (usually punk or indie in style) and cabaret performances (usually with some form of queer political undertone or genderfuck theme) to be interspersed between EDM sets.

Other extramusical activities that commonly occur in queer scene spaces include alcohol and other drug[5] consumption, and flamboyant displays of costuming and sexual play. The latter may occur in the context of designated 'play rooms' or more discrete and darkened 'back rooms'. Depending on the nature and scope of the event, sexual play may be gender specific or segregated; however, in line with the body-positive politic of queer, middle-aged bodies of all shapes and sizes are highly visible in all spaces. In general, unencumbered sexual expression is a key scenic feature. While costuming is not ubiquitous, and casual dressing—in jeans and T-shirts, for example—is common, a significant number of people appear to use scene gatherings as opportunities for playfully dressing up.

Queer style cannot easily be pinned down to a specific 'look' in the same way that, for example, hip-hop clothing or skater-punk clothing could be described generically. Rather, the look of queerness is often drawn from an eclectic range of stylistic modalities, refusing classification within a singular logic of taste. Conscious, playful and politicized ways of 'doing' gender are the most identifiable indicators of queer style. In particular, gender ambiguity (androgyny), hyperbole (exaggerated femme qualities in women and exaggerated butch qualities in men) and artifice (female masculinity/male femininity) are common. However, the stylistic modalities that people draw upon in either gender-ambiguous or gender-hyperbolical dressing remain heterogeneous. These may include bondage or fetish-style attire, punk- and goth-style clothing, retro fashion, burlesque fashions or a variety of playfully ostentatious garments and accessories such as tutus, feather boas or sequins. Interestingly, post-youth subjects wore some of the most embellished attire I observed, and when I queried participants as to why this may be so, two dominant themes emerged. First, I was told that the older you get, the more 'tame' you are expected to dress—at work, for example—so opportunities for 'frocking-up' were highly valued. Second, the majority of my interviewees linked their queer ideologies to a more fluid, unrestricted and individually created sense of self-presentation in middle age: being queer means 'we already don't fit in' and with age comes greater confidence. Emily pointed out that queers care less about 'what other people think is important or by how people think we "should" behave . . . so when going out, our drivers are good music, dancing and dressing up'. Just as clothing serves as a common aestheticized way of 'troubling' gender (Butler 1990), for the post-youth subject, music, dancing and dressing up can also be read as aestheticized ways of 'troubling' age.

Scene gatherings most commonly occur at night.[6] Beginning around 9.00 p.m., an event may officially cease at 3.00 a.m. or 5.00 a.m., depending on venue licensing laws—if indeed the event is being run legally. Illegally run events are not uncommon, as queer scenes often exhibit a strong do-it-yourself (DIY) ethos. In contrast to the increasing commercialization of the gay public sphere, queer scene events are often staged in available-for-hire or fly-by-night spaces rather than commercially branded, fixed venues. For example, a series of scene-specific party nights—perhaps staged monthly or quarterly—may be held at different venues around the city.[7] In the Brisbane scene, DIY not-for-profit party nights facilitate social gatherings in decidedly queer spaces outside of the commercially operated mainstream lesbian and gay clubs, which usually are open nightly and attract large crowds every weekend. While official start/end times are usually prescribed on event flyers, gatherings may continue in the form of an 'after-party' or 'recovery', whereby friendship groups within the scene may splinter off to different smaller locations (sometimes a private residence, a day club or a form of hijacked public space) and continue to party well into the next day. A single night out could constitute anywhere from six to twelve hours of continuous partying.

These features of the scene—loud and typically 'youthful' styles of music, dancing, queer performances, drug-taking, public sexual activity, playful costuming, the late start time, the length of a typical night out, and the degree of commitment that it takes to stage DIY events—are indeed incongruous with both hetero- and homonormativities of middle age. While not all ageing scene participants may partake in all of these activities, most will partake in at least a few—with appreciation for the music and late start times of events being requisites for even the most basic level of participation. Of my interview sample, five respondents played key roles in organizing scene events. As Emily pointed out, 'A lot of the big or good event producers in our queer world are [aged] somewhere between forty and sixty-five . . . so even if the crowd is younger, the faces behind it aren't.' Post-youth participation in such activities can thus be said to 'trouble' or 'queer' a subject's age-specific cultural intelligibility. When I asked Roger whether he thought that queer identification and scene participation somehow disrupted age-related normativities, he said:

> Yeah definitely, you know the normal path like you have fun when you're at university or whatever and then you stop listening to new music, stop going to clubs, settle down and get into a sensible relationship, have children, focus on your career whatever . . . there are some people in my social group who formerly would be party people who now have kind of withdrawn into that world of parenting, career, whatever and who don't go out and party much anymore and I would say, yeah, those people don't identify with queer culture. But there's also lots of gay and lesbian people who do the same thing, who get into their little house buying, child minding, suburban, career, or whatever existence and they classify themselves and being too old to behave that way anymore, whatever that is. It's all just a matter of attitude really.

Appreciating the variety of lifestyle expressions the scene accommodates, participation is not something Roger—or indeed any of my other interviewees—is willing to forsake with age: 'I might as well pull the plug if I get that old,' said Roger. In Bennett's (2006) previously cited discussion of ageing punks, he found that punk status could be maintained through modifying aesthetic and discursive practices in ways more conducive to adulthood. Here, however, queers do not appear to be modulating their participation to suit adulthood; instead, they are modulating their perceptions of adulthood to ensure continuing participation.

For Jacob, participation in this scene means that he is afforded the opportunity to perform his middle-aged sexual identity in a way that is most pleasing to him, while also acknowledging that outside the scene, and especially in the highly corporate and conservative world in which he works, 'there's always going to be people who think it's appalling'. He goes on to say that his main reasons for continuing participation are

> largely to do with music and dancing . . . music gives me a chance to dance and be silly and have fun . . . I don't think in my head that you're ever too old . . . If I decide I want to wear high heels and make-up and mismatched fucked up clothes or a dress or something it's totally fine. Yeah, there's no level of discomfort about that at all, well maybe a little bit getting to the venue and home again [he laughs] . . . Then there's drug taking, there's being excessive, there's sexual activity with pretty much whoever . . . you can just modulate all that however you like, I don't think you're judged for that stuff.

For Jemima, scene participation also remains highly desirable in middle age: 'I don't feel too old . . . and I don't actually envisage myself being left behind.' While she admits that frequency of participation fluctuates more in middle age, depending on career demands and physical health, she says that being connected to 'queer life' is essential for her. She explains:

> I had a complete rejection from my family when I came out, this was many years ago, so queer life was my family so I've always been quite connected to it and it just keeps me going, like, I need my queer fix, I work in such a straight world generally that I tend to just want to have that identification met on some level.

While all of my interviewees identified as same-sex attracted (either exclusively or mostly) in terms of the scene generally, the specificities of sexual orientation cannot so easily be delineated: sexual affections between men and women, while less frequent, are still noticeable. As a result, it is becoming more common for scene members to use descriptors such as 'bent', 'kooky', 'quirky', 'queer and alternative', 'alternative straight' and 'queer straight'. These queer tropes signal a range of practices and lifestyles—both homosexual and heterosexual—that defy the conventions of hetero- and homo-normativities. Moreover, as we can see from the comments below, heterosexuality in queer scene contexts appears to intersect to some degree with ageing subjectivities. Roger and Emily, respectively, remark:

> There is actually quite a lot of older heterosexual people who hang out with us now because of that non-judgemental thing and they know that we know how to have a good time . . . regardless of their sexual preferences, regardless of their age, as long as they are into it and respectful . . . no dickheads is the only rule, so I think that alternative straight people have worked out how to do this stuff properly.
>
> I think the straighter the environment, like 'straight' in inverted commas, so not necessarily heterosexual, but straight thinking . . . you know, the quirky bent end versus the straight conservative end, I think the straighter you are, the more you age, the faster you age and the more judgments that are placed on you.

Where older heterosexual people who are 'more alternative' or 'sexually explorative', or desire this type of post-youth scene engagement, feel impeded by ageing normativities, the queer ideologies and politics imbued in this scenic composition appear to offer a degree of relief, inspiring alternative presentations of ageing selves for multiple sexual subjectivities—especially those that are nonreproductive.

In the previous quotes, as in much of the interview data generally, it is evident that scene members recognize the discursive construct of age and the judgements ascribed to a subject on the grounds of ageing intelligibility/unintelligibility or proper/improper subjectivity. Given that heteronormativity 'is produced in almost every aspect of the forms and arrangements of social life' (Berlant and Warner 1998: 554), including temporal arrangements, queer scenes stridently aim to operate according to differently emphasized notions of lifetime within which alternative narratives of ageing unfold. Instead of labelling these belated modalities of cultural activity 'arrested development', in which adult responsibility is diluted or disregarded (see Calcutt 1998), might it not be better to conceptualize queer scenic activity as simply another way of doing adulthood—a way that problematizes clear youth/adult demarcations? Ageing is also recognized by my interviewees as a somewhat fluid and performative assemblage that can 'appear' more or less exaggerated, faster or slower, depending on how one behaves. Beyond the scene context, ageing may appear more exaggerated or accelerated, yet from within, 'perceptions of middle age are not so acute', as one interviewee said, suggesting that even in middle age, the 'queer world' affords her the space for youthful self-presentation that work time and family time do not. These post-youth displays of scene commitment profoundly illustrate how age—like gender and sexuality—is in effect constituted through flexible actions and 'choices' that may or may not be congruent with the physical signs that mark a body beyond youth.

Conclusion

In this chapter, I have examined the significance of musical and extramusical scene-specific activities in the performative assemblage of post-youth identities. Specifically, I have related this to a Brisbane-based queer scene and a selection of subjects whose post-youth sociosexual identities are intrinsically tied to and stylistically mediated within this scene context. The case presented here further cements the argument that music scene participation and accompanying articulations of style can retain critical currency in adulthood. Building upon this, I have demonstrated the potential for scene participation to effect alternative visions of ageing selves. Grounded in queer theory, the argument has been that post-youth scene participation significantly troubles normative temporalities of ageing in both hetero- and homo-normative contexts, calling into question the stability and authority that adulthood acquires through its distinction from the time of youth. Ultimately, just as drag performance has the potential to weaken and destabilize the category of gender (Butler 1990), the scene-specific performances of ageing queer identities discussed in this chapter potentially weaken the supposedly linear correlation between one's physically ageing body and intelligible social age. At the same time, this case study's focus on the post-youth queer subject forces a reconsideration of the master narratives of youth and reproductive sexualities in the context of studies of youth culture and popular music studies generally.

In conclusion, it would be remiss of me not to highlight the shortcomings of this study, specifically in relation to its inability to accommodate a critical analysis of ethnic and/or racial post-youth queer scene identification beyond whiteness. While one person out of the ten respondents I cited in this article was Indigenous Australian, that person did not mention Aboriginality in the context of our discussions on ageing. My case study, as with the existing literature that I reviewed, fails to ask how race and ethnicity are imbricated in forms of post-youth scene participation, and whether being from a different racial or ethnic background problematizes post-youth scene participation in other ways not previously considered. As such, this is a valuable area for future investigation.

–3–

Ageing Rave Women's Post-Scene Narratives

Julie Gregory

Introduction

In keeping with McRobbie and Garber's (1976) concerns about an absence of female voices in the field of youth culture research, contemporary scholars argue that women's experiences continue to be conspicuously absent from research into club cultures. Indeed, with few exceptions (e.g. Gregory 2007; Hutton 2006), most recent scholarship fails to take as its central focus women's specifically located experiences within rave scenes. According to some researchers, this means that scholars are missing important opportunities to consider how 'the stories raving women tell say a lot about emerging fictions of femininity'—how, in other words, they conform to or challenge dominant understandings about 'proper' gender roles (Pini 2001: 15). To this, one might add that by not exploring women's narratives as they move beyond the category of 'active rave participant', researchers also gloss over important questions about possible convergences between (non)rave participation and normative meanings attributed to ageing bodies.

The purpose of this chapter is to fill in some of these gaps by offering insights into the negotiation processes undertaken by ten women as they shift further away from active participation within Toronto's rave scene.[1] The findings presented in this chapter reveal that dominant gender- and age-related discourses figure centrally in these women's accounts of their bodies as no longer being appropriate sites for displaying an active 'raver' identity (through particular drug use, dancing and clothing tastes, for example). Relinquishment of their raver identities notwithstanding, given interviewees' descriptions of rave participation as providing them with religious-like experiences (see also St John 2004), it should not be surprising that they described these experiences as shaping them in ways that extend beyond their years of active rave participation.

In what follows, these findings are teased out in four sections. First, an overview is offered of the methods used to garner interviews. Next, I explore some of the ways in which discursive and material understandings of age and gender converged with interviewees' accounts of their shifting body images and their stated reasons for renouncing their active raver identities. Third, I draw attention to interviewees' constructions of their post rave scene lives as shaped by sensibilities they attribute to their experiences within Toronto's rave scene. The chapter concludes with a review of my main findings and an exploration of what these suggest in terms of possible directions for future research.

Meeting the Women

The findings presented in this chapter are based on semistructured, face-to-face interviews with women who met the following criteria: they were at least 25 years old; they began participating in Toronto's rave scene between 1994 and 2000; they attended these events regularly for at least one year; and they no longer identified as active 'ravers' (see Table 3.1).[2] The purpose of these conversations was to provide interviewees with a space within which to narrate their experiences with becoming, being and ceasing to be active rave-goers. Special attention was paid to the ways in which interviewees depicted age and gender as influencing these movements.

I used opportunistic sampling methods to generate this sample of interviewees. Logistically, this meant that as a self-identified ex-raver, I first approached women in my social circle who met the research criteria and invited them to participate in the study and/or to put me in contact with other women who might be interested in participating. This sampling method allowed me to interview ten women. At the time of their respective interviews, research participants ranged in age from 25 to 39. My

Table 3.1. Interviewees' ages and years of active rave participation

Pseudonym	*Age at interview*	*Age introduced*	*Year first attended*	*Years most active*	*Year last attended*
Cosmic	29	18	1994	1994–2000	2002
Curious	25	16	1996	1997–1998	1999
Grrrl	28	16	1993	1993–1995	1997
Kickin'	25	15	1995	1999–2000	2000
Molly	26	18	1998	1998–1999	2001
Mystic	27	16	1994	1994–1998	2005
Penelope	29	19	1996	1998–1999	2004
Pink	26	18	1998	1999–2003	2005
Princess	28	17	1995	1999–2001	2005
Serendipity	39	33	2000	2000–2002	2005

conversations with these women took place between January and April 2006, and ranged in duration from one to three hours. Due to the highly sensitive topics around which these discussions revolved (e.g. drug use, personal relationships), and because in many cases I was talking with people who had never met me before, interviews took place in locations research participants considered safe spaces. Following this same rationale, each interviewee was asked to choose a pseudonym which she felt suitably captured some aspect of her past active rave participation; it is these pseudonyms that I use in this chapter to identify interviewees.

It is important to note that my chosen sampling criteria—that is, women over the age of 25 who participated in Toronto's rave scene between 1994 and 2000 for at least one year, but who no longer self-identified as active ravers—had a number of consequences. First, these relatively broad criteria allowed for a great deal of variation in terms of the timeframes and ages discussed by each woman: at one end of this spectrum is Kickin', who attended her first event in 1995 at the age of 15, while at the other end is Serendipity, who went to her first rave event in 2000 when she was 33 years old. Second, these research criteria also provided for a great deal of variation in terms of the length of each woman's rave participation, with four women (Curious, Kickin', Molly and Penelope) attending raves regularly for one year, three (Grrrl, Princess and Serendipity) attending regularly for two years, two (Mystic and Pink) for four years and one interviewee (Cosmic) attending rave events regularly for six years. Third, my chosen sampling criteria made room for interviewees to interpret their positions as 'no longer active ravers' in varied ways, with one interviewee claiming not to have attended a rave event in the last eight years (Grrrl) while others (Mystic, Pink, Princess and Serendipity) were still going to 'rave-type' events, but on a less regular basis than they had in the past. Finally, I chose to employ purposive sampling techniques as a way to garner data that offered unique insights into women's reasons for relinquishing their active raver identities and the extents to which they understand their post rave scene selves as shaped by their past active rave participation. This advantage notwithstanding, this choice limits the generalizability of my conclusions to broader discussions about the (non)compatibility of rave participation and getting older.

In light of this final point, it is important to note that in the larger study from which the present discussion is drawn, interviewees spoke extensively about their movements into and within Toronto's rave scene (Gregory 2007). For the purposes of this chapter, these aspects of interviewees' narratives will figure in the present discussion only insofar as they intersect with their stated reasons for abandoning their active raver identities. For example, the next section begins with a discussion of the central roles played by clothing, dancing and drugs vis-à-vis interviewees' active raver identities. As this analysis develops, however, it becomes clear that these same factors directly informed interviewees' sentiments that their ageing female bodies eventually would no longer be conducive to active rave participation.

Displaying an Active Raver Identity

As interviewees discussed their past identifications as active ravers, they consistently spoke about the enjoyment they derived from exhibiting this identity through particular clothing, dancing and drugs choices. For example, when probed about their typical rave experiences from beginning to end, Cosmic, Curious, Grrrl, Kickin', Mystic and Pink enthusiastically recounted pre-event rituals during which they and fellow female rave-goers got together to choose hairstyles and makeup colours and/or make 'rave clothes, quote unquote', as Pink articulated it. This finding reveals that preparation for rave events involves physically displaying a raver image and, at least for these women, this is understood as an integral component of active rave participation and their accompanying raver identities (see also Thomas 2003).

In keeping with these insights, interviewees noted the converging roles their clothing, music and drug preferences played in relation to their identifications as not only active ravers, but also as particular kinds of active ravers. This tendency is illustrated, for example, by Molly's claim that crystal methamphetamine 'was definitely [her] drug of choice'. She continued:

> The clothes were probably the biggest thing for me. I got way into the fashion and started dressing like a raver and wanted a piercing of course. Plus you always distinguish different groups of ravers . . . I identified myself with the hard-house ravers, the real dancers that liked to get out and move.

Again, as these insights indicate, a number of interviewees constructed their past active rave identifications not only through particular consumption practices, but also with reference to the ways these choices differentiated them from other rave-goers. This finding is not surprising, given that it buttresses other depictions of rave spaces/participants as stratified along music, drug and clothing axes (e.g. Malbon 1998; McCaughan et al. 2005).

What does emerge as novel, however, is the extent to which interviewees represented their identifications as active ravers as being linked directly to their particularly aged and gendered bodies. For example, Molly constructed her past active rave participation as follows:

> I mean I was 5 feet 1 [inch] and I was 80 pounds soaking wet and when I went to raves, guys loved that. *Ravers love little girls*. It's a little sick, I know, but I never had so much attention that I did at raves. I've always had a knack for getting things I want 'cause I sweet talk people, but at raves it was so easy; we went to a rave once with no money! We just had a bag full of jellybeans, and I . . . managed to talk us into some drugs and into some water bottles, just with jellybeans, and . . . it's just 'cause it was me, and people liked little girls, and then of course when I told them my age, then they felt like psychos 'cause I wasn't that young.

Here Molly represents the confidence she once felt embodying an active raver identity as contingent on her being a 'little girl'—which she conflates with having a thin *and* young (looking) female body.

As these comments foreshadow, just as interviewees understood their younger female bodies as particularly well-suited to an active raver identity, so too they had 'come to recognise their own ageing through [perceived] changes in bodily appearance' (Twigg 2007: 290). To substantiate this insight, consider that when asked if she thought she would ever go to another 'rave-type' event, Molly responded:[3]

> No . . . It wouldn't be the same . . . [because] I'm older. I have responsibilities now, and I don't feel good enough to go to a rave . . . I don't feel like I'm fit enough to stay up that late. I don't have enough confidence in myself . . . I don't feel pretty enough to go out there and . . . dance in a tiny tank top.

By suggesting that she no longer feels 'pretty enough', as an older woman, to embody an active raver identity—via 'danc[ing] in a tiny tank top', for example—Molly redraws attention to claims that clothing choices 'play a vital . . . role in the constitution [and display] of social difference' (Twigg 2007: 291). Put differently, following Molly's comments, age, gender, clothing choices and body image converge in ways that shape people's identification as active ravers *and* their feelings that, eventually, they become too old for active rave participation.

Building on this analysis, other interviewees framed eventual relinquishment of their active raver identities in terms of disjunctures they felt existed between active rave participation within Toronto's rave scene and increasing senses of responsibility outside rave spaces. After glancing at her pregnant belly, for example, Penelope said she was not likely to go to another 'rave-type' event. She clarified that 'there were a few people we knew who did have kids and would still go out and party and we never thought that was a responsible thing to do'. This led Penelope to conclude that rave participation is 'just one of those things that ends when you have kids'. In this sense, Penelope provides at least partial support for claims that participation in and/or appreciation for popular cultural phenomena 'shift place in one's salience hierarchy due to the emergence of competing priorities' (Harrington and Bielby 2010: 438). More specifically, Penelope's insights redirect attention to the finding that dominant discourses which dictate age- and gender-(in)appropriate behaviours infused interviewees' explanations for why they stopped attending rave events on a regular basis.

Grrrl's response when asked how/why she became less active in Toronto's rave scene is particularly telling in this context:

> We're older now, we should focus on our career, savings . . . and if you're doing drugs at this age, it's like what are you doing? . . . You're supposed to be moving forward with yourself, not backwards. Like when you're in your early teens, you're trying to find yourself, you can fuck up and do that shit, but you should be a little more mature by this age and . . . it's [drug use] harder on our body too at this stage; you don't snap back as fast from it.

Leaving aside discussions about whether or not drug use is a necessary component of active rave participation, the above insights speak to a number of interesting and related issues vis-à-vis (non)rave participation, age, gender and body image.

First, Grrrl frames the purported necessity of her current nonactive raver identity in terms of older women's bodies being particularly sensitive to the effects of drugs. Interestingly, this was a theme that emerged—albeit in various guises—from my conversations with all ten women. For example, compared with Grrrl's representation of rave participation (particularly drug use) as having immediate and negative corporal consequences for (older) women, Cosmic represented continued rave participation as having a direct impact on women's physical appearance. As she said, she was unlikely to attend another 'rave-type' event 'cause women age terribly bad[ly] in the scene'. When I asked her to elaborate, Cosmic rationalized, 'I think that drugs have a more ageing effect on women and prettier women come and the guys get older and they bring in the young pretty girls . . . Women don't look good older in the scene.' Other interviewees reproduced similar constructions of older women's rave participation as particularly risky when they voiced concerns that their rave-related drug use—as compared with that of their male partners—has directly affected their abilities to conceive (healthy) babies in the future (Gregory 2009; see also Gregory 2010). Together, these comments indicate that as they got older, interviewees began to understand—if not experience—active rave participation in terms of exclusionary discourses that position certain cultural activities (rave participation, drug use and particular clothing styles) as *necessarily* the purview of certain types of people (young, fit, nonfemale bodies).

At the same time, interviewees' rationalizations for ceasing active rave participation might be read as buttressing 'commonsense' beliefs that people who disregard 'the equation of legitimate popular music fandom [and related activities] with youth . . . are . . . social misfits, hankering after a life in which they no longer have any rightful place' (Bennett 2006: 221). Indeed, as already suggested, the overwhelming majority of interviewees constructed active rave participation as a phase that people—particularly older women—must relinquish if they were to avoid having what one interviewee (Pink) referred to as 'a sad life'—one without a life partner, children and/or a 'good job', for example (see also Gregory 2009; Pini 2001). In this sense, interviewees' narratives about their shifting relationships with Toronto's rave scene can be read as relying on and reproducing (lay and academic) denials that not everybody experiences active rave participation and 'adulthood' as incompatible (Gregory 2012).

Serendipity's story emerges as particularly interesting in this regard. Contrary to other interviewees' accounts of renouncing their active raver identities because of their older ages, Serendipity spoke extensively to the difficulties she had reconciling her relative position as 'too old' (and a mother) with her entry *into* Toronto's rave scene. Specifically, she told me that as a 33-year-old attending her first rave events, she felt very insecure knowing that she was much older than the overwhelming majority of other rave attendees. Serendipity also described these experiences as 'extremely special', and thus as superseding her apprehensions that she 'was probably too old to be there'.

This is not to say that the anxieties she felt as an older raver disappeared. On the contrary, and buttressing Molly's claim that as a young (looking) woman she had special privileges within rave spaces, time and again during our conversation Serendipity recalled feeling 'extremely lucky' that she 'looked really young' because, she reasoned, this meant other rave-goers likely could not 'guess [her] secret' (that she was likely much older than them). In this same context, Serendipity told me that she often took extra precautions to help ensure that she would continue to 'pass' as a (young-looking) raver. For example, she said she dressed 'androgynously' (e.g. in baggy jeans and a T-shirt) for rave events so as not to accent her older female body. In addition, despite her clear embarrassment, Serendipity admitted that she once asked a younger co-worker of hers, who she said 'was a total raver girl', to teach her how to 'raver dance'.[4] In these ways, Serendipity's experiences seem to challenge other interviewees' understandings of an older chronological age as necessarily coinciding with a biologically older, less resilient body—one that is thus interpreted as out of place within rave spaces. Stated more theoretically, Serendipity's narrative makes room for understanding age—like gender—as performative insofar as it can be manipulated through particular consumption practices, for example (see Butler 1990, 1993).

While the importance of this interpretation should not be denied, it is notable that Serendipity complicated this discursively informed depiction of her active raver identification as she began to reflect on her reasons for leaving Toronto's rave scene. Specifically, when asked why she put an end to her active rave participation, Serendipity explained:

> I quit smoking. Yeah, that, I kind of went through this misery of quitting smoking, didn't wanna, and I think that I'd been thinking about it for quite a long time that I don't want to be doing this anymore, really about that safety issue for me: I'm a mom, I shouldn't be doing this. I don't know why I put that on myself because people will . . . like why can't I be doing this, but it's really because they're mystery pills. If it wasn't, I don't think I'd have, if I knew what I was getting, yeah, I'd probably go more. It's just that, it's the only thing that I like to do; I don't like drinking, I don't like weed, I tried cocaine a couple of times, I don't like it, you know. I like this, this is good, but it's scary, too.

Again, analyzing these comments closely reveals that Serendipity's decision was informed by cultural/discursive as well as biological/material readings of her body.

More poignantly, these comments make clear the extent to which Serendipity felt torn between health-related concerns that rely on understandings of the body as a biological entity (such as her use of nicotine and knowledge that people can never be sure about the exact concoction of chemicals they ingest when they take illicit narcotics) on the one hand, and acceptance that her aged and gendered body is made meaningful via discourses that construct rave participation, drug use and motherhood as incompatible on the other hand. They point, in other words, to the finding that (some) people's decisions about rave scene membership are fraught with 'tension[s] between personal and structural identity and the strategies people use to live and develop in circumstances not of their own choosing' (Biggs 2004: 46). Following this reading, Serendipity's experiences as an older raver do not contradict other research participants' claims that eventually they got too old for rave; instead, they underscore the finding that interviewees' decisions to withdraw from Toronto's rave scene were as much a by-product of deeply entrenched age and gender-related discourses as they were about their experiences of 'the effects of the materiality of the body in age' (Woodward 2006: 186).

Ultimately, all ten interviewees abandoned their active raver identities. Again, the most common reason for this was that as they got older they no longer understood (or experienced) their bodies as conducive to embodiment of an (ideal) raver identity. This is not to suggest, however, that interviewees constructed themselves as completely disconnected from rave. As the findings presented in the next section illustrate, interviewees' narrations of their shifting relationships to Toronto's rave scene include understandings of their post rave scene lives as directly shaped by their past rave-related experiences.

Embodying a Raver Sensibility

Notwithstanding research participants' tendencies to describe themselves as no longer 'pretty enough' to attend rave events, many of them constructed their past rave participation as mediating their current positive self-image. Mystic, for example, had the following to say when I asked what rave had meant to her:

> Raves really brought me out of my shell . . . I was always kind of a really shy person. I always had a lot of friends . . . but I was shy with meeting new people . . . it would have to be someone else making the effort to meet me. And um, and I feel like raves, like being part of this really accepting community, really made me a more open and extroverted person.

Here Mystic describes her experiences within Toronto's rave scene as having an especially positive impact on her self-esteem and her ability to feel connected to a larger community of people—a reading that buttresses both other interviewees' narratives and scholarly claims that rave participation fosters intensely spiritual feelings of peace, love, unity and respect, or PLUR (Hutson 2000).

Kickin' expanded on this insight when she rationalized that her past active rave participation—particularly the dancing aspect—increased her self-confidence and that this heightened self-image had translated into greater respect for other people. Specifically, Kickin' told me that before she entered Toronto's rave scene she was 'snobby', but that that began to change as she met fellow rave-goers who, she said, were 'so accepting . . . so loving and . . . just happy'. She concluded, 'You know, growing up with parents who are just so cynical and so judgemental of everybody, I was worried that I was going to turn out like that, but I think that [my participation in Toronto's rave] scene completely made me not be that way.' Again, Mystic's and Kickin's claims reveal that at the same time that many interviewees framed their cessation of active rave participation with reference to negative stereotypes vis-à-vis older women's body images and capacities, they depicted this same participation as having lasting positive impacts on their sense of self.

This reading resonates with Hunt and Evans's (2008) finding that use of the drug Ecstasy (aka 3,4-methylenedioxymethylamphetamine) at electronic dance music events 'serve[s] as a catalyst for creating new possibilities in how [respondents] perceived and related to themselves and the world around them' (342). Like the women with whom I spoke, the 276 people Hunt and Evans interviewed said that Ecstasy use cultivated feelings of confidence and gregariousness amongst clubbers. Following these insights, there is a need for research that explores how (and whether) people's rave-related experiences shape them in ways that extend beyond the confines of individual rave scenes and/or beyond their years of active rave participation.

In relation to this opening in the related literature, it is worth considering that Cosmic, Penelope and Serendipity told me that their past rave-related experiences had directly contributed to their nonjudgemental attitudes in their respective positions as a case worker at a local shelter for street youth, a registered nurse and an outreach therapist for 'juvenile delinquents'. For example, Serendipity said her rave-related use of Ecstasy facilitated such remarkably profound experiences that they prompted her to become a major proponent of harm-reduction education.[5] She highlighted the extent of her appreciation for this approach and her related marginalized position at her place of employment when she said 'at work, *I am* harm reduction'.

In a similar vein, Penelope told me that it was her experiences within Toronto's rave scene that had caused her to be 'a lot more non-judgmental'. She elaborated:

> I used to be fairly judgmental about it and, you know, bought into the . . . perception of people who used drugs, that they all became, you know, crack addicts and have to hit bottom, so I think that has definitely changed and it affects the way that I interact with patients here. We do a young mothers program and I'm not nearly as judgemental I think as other nurses if they've used marijuana or they've used Ecstasy . . . in the past whereas a lot of [other] nurses think, 'Oh well, they must be hardcore addicts if they've used Ecstasy.'

Penelope concluded this portion of our conversation with an explanation that while she rarely ingested drugs as a rave attendee, any stereotypes she had about drug use(rs) faded as she met people within the scene who 'still had regular, responsible, normal lives and who were functioning adults despite their recreational use of drugs'.

As a final example of links interviewees drew between their past active participation and their current self-understandings, consider the following insights Cosmic offered during the debriefing period that followed her interview:

> It was nice to actually do some self-reflection and look at how it's shaped my life today. I never realized, like I knew it had a big impact on me as a . . . on my self-concept, but I didn't realize that . . . has actually persuaded me to go into my field. As well as giving me to the tools to have the self-confidence to excel . . . At my work I'm the youngest case manager there and one with the heaviest caseloads, which I can handle . . . because I think . . . my [past rave] experience . . . has given me the knowledge base, the emotional stability and just the confidence to do it.

Here, Cosmic draws together findings that past active rave participation mediates interviewees' positive self-image and nonjudgmental attitudes when working with marginalized populations. More than this, when read in conjunction with the findings presented in the previous section of this chapter, Cosmic's reflections reveal that while interviewees eventually felt out of place within rave spaces, they also understood their past identifications as ravers as facilitating their feelings of being especially well-placed in other social arenas.

Building further on Hunt and Evans's (2008) discussion of the possible lasting influences that Ecstasy use may have in people's lives, Cosmic (like other interviewees) spoke extensively about the positive effects she experienced from dancing at rave events—intoxicated or not—with hundreds, sometimes thousands of people who shared her love of electronic music. That said, these aspects of her narrative were interspersed with accounts of some of her more painful rave-related experiences (such as losing a close friend to a drug overdose and being sexually harassed). When asked about these decidedly more troublesome experiences, Cosmic was adamant that it was the *culmination* of her positive and negative experiences within Toronto's rave scene that allowed her to connect with clients in ways she felt were unavailable to her co-workers.

As this point suggests, interviewees not only understood their past rave-related experiences as shaping their abilities to be especially compassionate employees; they also constructed these experiences as encouraging them to become more socially aware and politically active, thereby supporting claims that participation in popular cultural scenes can inspire people to become committed to (even 'small' acts) of social change (see Mattson 2001).

In this context, Serendipity said that in addition to shaping the ways she understood her role as a paid therapist, her past rave-related experiences had compelled her to begin volunteer work with TRIP!, a Toronto-based organization that provides members of local dance scenes with safe drug and sex information.[6] In keeping with the current discussion, Serendipity constructed this latter work as a way for her to spread her belief in the importance of empowering people to 'make their own choices' regarding drug use and as a venue through which she could remain 'somewhat connected to [Toronto's rave] scene'.

Grrrl's experiences further substantiate this challenge to readings of rave participation as having little significance beyond fleeting expressions of and resistance to social norms (Thornton 1995). Specifically, towards the end of her interview, Grrrl told me that while she currently worked as a full-time manager at a local advertising agency, she continued to be connected to Toronto's electronic dance music scene via her part-time employment as a DJ. In talking about her transition from a young, dedicated raver to a more mature, part-time DJ, Grrrl (like other interviewees) described a number of traumatic experiences she had within the scene—including, but not limited to, sexual assault, and gender- and age-based discrimination (Gregory 2007). As opposed to interpreting these experiences as unequivocal confirmation of media-led representations of rave spaces as uniquely dangerous, she—like Cosmic—said they afforded her with insights she would not have gained otherwise. In addition to providing her with 'an awaking for . . . how much of a struggle women still have', Grrrl told me her experiences within Toronto's rave scene had led her 'to feel empowered and to [want to] help other people'. She continued:

> Whenever I have a platform, I . . . give other females a voice to do it as well. Females and males, a lot of my friends are artists and they are really big introverts, you know, and they're really amazing musicians and if I can help them out by getting them out there . . . then that's amazing.

Again, contra to claims that rave has very little to do with 'real' politics and in keeping with feminist-inspired calls to appreciate the inseparability of capital 'P' and small 'p' politics (Hanisch 1971), interviewees' narratives reveal rave participation as *at least politically significant* insofar as it continues to shape interviewees' 'way[s] of being and interacting' in the world (Hunt and Evans 2008: 345). Following interviewees' narratives, it is clear that their current lives were shaped in specific and meaningful ways by the feelings and perceptions (that is, sensibilities) that they acquired during their periods of active participation within Toronto's rave scene.

Looking Back, Moving Forward

The findings presented in this chapter are commensurate with claims that group identification can have residual meanings for people 'beyond the time an identity is abdicated' (Torkelson 2010: 268; see also Ebaugh 1988). They also resonate with Bennett's (2006) finding that, at least for the older fans with whom he spoke, punk is best understood as a *lifestyle*, 'a set of beliefs and practices that have become so ingrained in the individual that they do not need to be dramatically reinforced through the more striking visual displays of commitment engaged in by younger punks' (226). Admittedly, rather than suggesting that rave has become an '*ingrained*' aspect of interviewees' lives, the findings presented in this chapter suggest that the meanings interviewees attribute to rave will continue to morph as they confront new discourses and experiences.

In relation to this claim, it is useful to recall that as interviewees reflected on relationships with Toronto's rave scene, they recounted a process whereby discourses of youthful pleasure that informed the ways they embodied their active identifications via particular clothing, music and drugs choices lost their saliency as discourses of adult (ir)responsibility began to merge with material and discursive readings of their older bodies. At the same time, interviewees' reflections on the current positions as older ex-ravers reveal some of the ways in which they have incorporated aspects of their past identifications as ravers into their current lives and senses of self. This latter insight is significant insofar as it complicates readings of interviewees as unwittingly reproducing commonsense understandings of age- and gender-appropriate behaviours by making room for interpreting their past active rave participation as providing them with opportunities to challenge related stereotypes in their social roles outside of Toronto's rave scene.

Again following these findings, instead of substantiating claims that people 'superimpose attributes of the self, their beliefs and value systems . . . onto the object of fandom' (and vice versa), the findings presented in this chapter might more fruitfully be interpreted as pointing to intersections between the fluidity of people's experiences and the meanings they attach to those experiences (Sandvoss 2005: 104). They should be read, in other words, as highlighting the extent to which the ways people make sense of and narrate their lives are intimately tied to their material and discursive experiences as particularly located bodies. Importantly, as those experiences and bodies change, so too will the stories they tell.

In light of this reading, the findings presented in this chapter suggest a need for more research into the negotiation processes undertaken by women who must contend with normative beliefs that older female bodies are particularly ill-suited to active rave participation. At the same time—and especially in light of Serendipity's experiences—the present discussion might compel researchers to continue to explore the extent to which older women are not so much absent from rave spaces as they are especially marginalized both within these spaces and within discussions of them. Regardless of the chosen path, given that interviewees' narratives as ageing ravers seem to correspond with shifts in their self-image, there is a clear need for further research into the ways (non)rave participation shapes and is shaped by people's materially and discursively informed understandings of their bodies.

The following questions might help to guide such research endeavours: how do relatively older people who remain in rave scenes negotiate discourses that situate them as not belonging? In what ways do people's rave-related consumption practices alter as they get older, and how do these changes impact other aspects of their identifications as ravers? How do older people who continue to self-identify as ravers embody and enact this identity as compared with their younger counterparts? Do older ravers embrace this aspect of their identities or do they engage tactics to hide it? In what ways do older ravers experience this identification as (not) commensurate with their other social roles? In addition to age and gender, how do other aspects of people's embodied identities shape their movements in and out of rave (and related) scenes?

Acknowledgements

Thank you to David Butz and Rebecca Raby for support with the research upon which this chapter is based, and to Andy Bennett, Paul Hodkinson and Samah Sabra for extensive feedback on earlier drafts of this chapter.

Part II
Constraints of the Ageing Body

–4–

'Each One Teach One': B-Boying and Ageing

Mary Fogarty

Introduction

In the late 1970s and early 1980s—the early days of 'breakdancing'—it was usual for participants to have retired from dancing by the age of 16 (Banes 1984, 1985). But now, some thirty years or so later, expert b-boys[1] and b-girls who began to dance in the earlier days are still competing. For example, Dolby D in London, who first made his reputation in the early 1980s, is still representing and battling[2] as both a rapper and a dancer. Similarly, some of the New York City b-boys and b-girls, including the well-known Ken Swift and Crazy Legs, who made their names in the early b-film cult classics such as *Beat Street* (1984), *Wild Style* (1982) and the notorious documentary *Style Wars* (1983), were representative of the ghetto youth movement known as hip-hop culture in the 1980s. They are now middle-aged pioneers concerned with the legacy of that culture and its preservation. The dance practice—or 'b-boying/b-girling' in the terminology preferred by dominant figures in the current scene—has now become well-established, with high-profile international competitions, theatre shows and documentary accounts of its history. Alongside the legendary dancers, older novice dancers are now also on the scene, such as Krazee Grandma from Sweden who began her practice in her sixties and represents to the fullest at international jams with the support of the scene behind her.

Hip-hop culture is made up of four elements: emceeing, breaking, graffiti and DJing.[3] Breaking, or b-boying/b-girling, is what the media once called 'breakdancing', although this was an umbrella term given incorrectly to dances (such as popping and locking) that originated on both the East and West Coasts of the United States. Breakdance became a popular name for the form during a critical moment of media interest in the mid-1980s, and has stuck with breaking ever since. In its original context of New York City, breaking was not only a dance style, but also a slang term to describe someone 'going off' or flipping out. When describing a dance style, the label b-boying/b-girling refers to the accumulative effect of a vocabulary of movement, done with authority and with an essence, style and attitude involving how one 'gets down' on the floor at parties. This style has specific genre conventions and cultural practices, which include the music—it is typically done to the instrumental break of a record. Steven Hager (1984: 103) describes hip hop as an 'experimental laboratory' that 'has created an art form so original and vital that black and Hispanic artists have gained access to the established New York art world for the first time'. This was as true for breakers in the dance world as it was for graffiti artists in the galleries at this time.

Since hip-hop culture consists of more than breaking, constraints resulting from the ageing of participants often lead to alternative modes of expression within the culture. In an interview, one Canadian dancer, now in his thirties, remarked:

> Breaking is a part of a culture with many art forms, and many of us don't just practise one. As we get older, have kids, go to jail, and have less time to train breaking, we tend to focus more on rapping and DJing, but I don't consider this to be falling off, as we are still repping the crew style and philosophy.[4]

Ageing dancers in international hip-hop dance scenes tend to change roles in a variety of ways, becoming DJs, coaches, mentors, judges or musicians. In doing so, they create the infrastructure that supports the development of an established aesthetic for the dance, and preserve dance styles through their actions and sustained involvement in aspects of the scene (Fogarty 2010b). A transformation of roles also changes the meaning of the scene: from rebellious youth culture to multigenerational tradition. Ageing dancers take on new positions that emphasize their increasing knowledge, even as their bodies begin to experience more and more aches and pains. These new roles have changed the international scene and filled out the infrastructures that support the dance style in dynamic ways, involving both dance practice and other musical involvement such as production and DJing. In hip-hop culture, dance and music are crucially related (see Fogarty 2010a), and I will argue later that this relationship reveals important facets of the ageing process of b-boys and b-girls.

This chapter has three major themes that coalesce around the sociological study of ageing in the context of a popular dance art form and practice. The first aspect to consider is the social and professional relationships between participants. These change as dancers become older and take on new roles in the scene. Here I use the analogy of the extended family to illustrate the dimensions of this practice, which are tied to intimate lifestyle choices. The second theme centres on musical mediation. Since dancers are performing their musical tastes, the ageing process can be examined through observing the relationships between different generations and the music to which they dance. Older dancers clearly shape the development of younger dancers, and their performance of taste is an integral part of this relationship, both in terms of aesthetic expression and in demonstrating competence in the culture's codes and conventions. The third theme is the manner in which age and knowledge about dance inform dance practice within the context of the culture. The literature about experts and novices reveals a key factor in distinctions between dancers and their status, reputation and prestige in breaking. Younger b-boys and b-girls may be expert enough to win competitions, but expertise is a complex peer category that is informed by involvement over hours, then months, then years. Here, the physicality of dance practice—the investment over time that is central to mastery and competence—is supplemented by a social hierarchy where authority is often attributed to the ageing body, even after the peak performance years have passed.

Background to the Research

This work is based on ethnographic field research, and some of the observations come from a broader research project on international breaking cultural practices that I undertook between 2007 and 2011. I interviewed b-boys and b-girls from Los Angeles, New York City, Toronto, Montreal, Berlin, Edinburgh, Glasgow and London, as well as danced with, competed against, judged and performed with b-boys and b-girls from various countries.

I further supplemented this project by conducting follow-up interviews with key participants who had mentioned ageing in passing during our conversations while I was conducting my initial research. These dancers were all between the ages of 27 and 45. Each dancer had been practising breaking for a minimum of seventeen years and a maximum of thirty-one years. The participants in this latter study were all male. I did approach several b-girls to answer follow-up interview questions via email or social networking sites, but did not receive any responses about this topic at that time. For this reason, I will supplement these observations with my own phenomenological experiences as a b-girl who has been breaking for thirteen years.

When conducting interviews, I often asked interviewees to recall their past experiences and memories as dancers. In what follows, occasionally I include quotes from b-boys who reflect on their personal histories. In these situations, I would also cross-check the verifiable aspects of their statements through a system of triangulation involving confirmation from event flyers or the memories of other dancers. There are limitations to this approach, as many psychologists have suggested that memories are not reliable (Chabris and Simons, 2010). However, memories are important in ageing (see Thompson, Itzin and Abendstern 1990; Chaney 1995). They tell us about the ways in which people construct the past, and within this practice they reveal the value systems highlighted by different generations of b-boys and b-girls, which are ripe for qualitative analysis.

The b-boys and b-girls in their late twenties and early thirties who I followed for ongoing interviews and participant observation were open to discussing their changing roles and activities. For example, I followed and danced with Karl 'Dyzee' Alba as he taught classes at the Street Dance Academy, trained neighbourhood kids in his apartment in Toronto, prepared and delivered a master class dance workshop in Montreal, taught at a community after-school program, competed in and won a b-boy crew battle at an outdoor hip-hop event (Under Pressure), took a Lindy Hop class and met with a project partner to develop a new judging system for breaking competitions.

Those in their forties were less interested in being interviewed on the specific topic of ageing. When I asked one informant of the larger study if I could interview him on this topic, he responded by saying in a disappointed tone, 'Oh thanks a lot, Mary.' It seems that some of those in a culture built on respecting those who have come before, and who are in the midst of trying to build a legacy, do not want to be associated explicitly with issues of ageing, preferring to legitimize associations with artistic practices and international influence. Having said that, a few of these older dancers offered up contributions informally, including thoughts on the question that now plagues their current and ongoing practice: 'How much longer can I do this?' Their desire to continue to participate was still present. But their body had begun to work against their ambitions. One respondent also explained that if I had told him at 16 that he would still be doing this dance at 40, he would have laughed. In this way, opportunities occasionally persist where ambitions of dancers are rather short-term in nature; this can be attributed to the desire for mentorship expressed by younger b-boys and b-girls. Younger dancers often encourage older dancers to continue to participate in the culture.

Extended Families

Thinking about ageing in relation to 'youth' practices—especially dance—involves questions about individual experiences of the ageing process, as well as about how social organizations develop categories for understanding ageing culturally. Early accounts of breaking, such as those of Banes (1985) and Leafloor (1988), used for their analyses theoretical frameworks informed by contemporary models of popular culture and youth studies. These studies emphasized the spectacular styles associated with so-called working-class youth 'subcultures'.

The early subcultural studies neglected ageing participants in their considerations because age was already a factor in the distinction they were making about the value of studying youth practices. Dick Hebdige (1979), for example, suggested that youth set themselves up not only against the dominant class, but also against their own parental (working-class) culture. Since early studies of 'breakdancing' were influenced by this model, writers such as Banes tended to position it in the context of a youth subculture, and thus as a 'youth-based' form of opposition to the parental culture. This overlooks the fact that older dancers have always been present and have often performed an important mentoring role for younger dancers. In the early days of breaking, the 'older' b-boys and b-girls tended to be two or three years older than the 'younger' dancers. The difference in the culture today is that sometimes the 'older' b-boys and b-girls are twenty to thirty years older. For participants within breaking culture, the implications of ageing come into focus when the centrality of 'crews' is examined, and this is vital for an understanding of how breaking culture works—what the dance means to participants and how central mentoring is to the process of learning a popular dance style.

Breaking has always revolved around crews—groups of individuals who formally decide to make their affiliation with each other known to the public. They name their crew, practise and compete together against other crews (known as 'battling'), and decide how to enact and negotiate their crew politics, involving issues of respect and seniority as well as dance competence. Each crew has a collective reputation based on the crew's abilities to dance and win at battles. Even from the onset of b-boy culture in the late 1970s and early 1980s, older b-boys and b-girls would often mentor younger ones. However, the age gap in this master–apprentice relationship has expanded substantially now that the practice is decades old, with some crews now containing members stretching over several generations.

An older b-boy, now well into his forties, described to me how he has begun to treat younger b-boys as if they were his sons. This raises the analogy of the 'family', a common articulation of close units of friends who collaborate on creative projects, similarly noted in interviews with local rock bands (Cohen, 1991) and b-boy crews. With the continuation of this form of dance, the family analogy has changed from a one-dimensional signifier of the closeness of peers who participate in a shared activity, as a crew, to what I argue are *extended families that are multigenerational in composition*.

Dyzee, a Canadian b-boy in his early thirties, recalls that he started dancing in 1994 when he had just turned 14 years old. During the summer, he saw breaking and immediately started to try it out. The first local crews he encountered in Toronto were Bag of Trix, Intrikit and Supernaturalz, as well as Crazy Legs and Bag of Trix on television. When he was first learning the dance, he recalls:

> I wasn't thinking I'm going to practise. It was just whenever I had time I would practise. I would just fool around all the time. Eventually I started getting good on my own. Plus, I used to go to all ages clubs and see breaking. I wouldn't even know what they did but I would see tricks and wonder how they did that and I would go home and try to make something up. Maybe it was similar and maybe it was different.

Many top performers with whom I spoke discussed the pleasure they experienced during their early encounters with dance, only later developing a disciplined and self-regulated training regime as part of their development.

During my research, I observed how at Dyzee's apartment, located in a high-rise building,[5] three or sometimes four boys from the block would knock on his door and wait. Dyzee would pull out a clipboard he kept inside his home with a list of moves that he had taught the boys so far. He would read out the moves and each of the boys would have to demonstrate the technique properly in the hallway. If each of the boys could do the moves adequately, then Dyzee would let them into his apartment where he taught them some more moves or showed them footage of his competitions or other dance events. Dyzee's living room transformed into a practice space. There is peer-to-peer learning inherent in this system because Dyzee did not let them into his apartment unless each of the boys could do the moves they had previously been taught. If there was one of them who did not have the move, it was the responsibility of the other boys to help him get it, so that they could all advance together. This system resembles a master–student model, straight out of a kung fu film or *Fight Club* (1999), and is—importantly—a free structure. Also, no one gets left behind—a radical pedagogical principle that involves not only master–student but also peer-to-peer teaching centred on collective efforts such as those found in the b-boy crew.

Poe One, another b-boy who is an advocate of the importance of passing along knowledge about the dance to new generations of dancers, is often heard saying 'each one teach one'. That saying—which has been found peppered throughout rap lyrics for the past thirty years—comes from the era of African American slavery, when blacks were discouraged and prevented from acquiring literary skills. Blacks and whites strove—often illegally—to teach slaves to read and write. For dancers, the phrase 'each one teach one' means that if you have had the opportunity to learn, you are obliged to teach another what you have learned. B-boys and b-girls have championed this phrase in an effort to keep b-boying alive in the spirit and essence of the dance form. The teaching strategy set up by Dyzee's system centres the students around a peer-to-peer learning approach that encourages teamwork, and this is one dynamic aspect about competitive popular dance practices that is generally misunderstood. There is an assumption that popular styles of dance such as breaking are competitive. This is obviously true, but this style is equally centred on mutual support and collaboration, as emphasized by the focus on crews.

Top-performing b-boys and b-girls, now aged between 30 and 40,[6] are quick to pull out their iPhones and show footage not of themselves dancing, but of the students they teach. In this way, they resemble proud parents, and their invested interests are clear. For example, Leon 'Vietnam' Carswell emphasized that an important aspect of his classroom teaching environment was making sure that students were having fun. Although the people who hire him want him to be more of an enforcer and disciplinarian, he thinks, as a foundation, the dance needs to be about pleasure and fun. In this sense, all of the top performers of breaking with whom I spoke understand the value of unstructured play in the development of skills—whether implicitly or explicitly.

Although many older dancers are paid to teach, all of them continued to teach other dancers informally at practices for free. To understand how the economy of the dance continues to persist within a model of exchange that is not centred on making money from other dancers, I use the analogy of the 'extended family'. Further, this also provides a way of thinking about the multigenerational and international constitution of crews more generally, and the relationships between individuals in a culture that has existed for more than forty years. This term will allow a focus on the sustained involvement that older dancers can have in the everyday lives of youth, informing their educational and ethical development as well as the parallels that are often drawn with elders in traditional societies. Family feuds, or tensions between crews, are also a dominant feature of this cultural practice, and speak to the social dimensions of performance within a competitive art world.

'Muscle Memory'

The involvement of older participants in hip-hop culture is often maintained through the efforts of youth participants. Older members of crews are encouraged by younger b-boys and b-girls to remain active, as their knowledge and expertise are valued. For an ageing dancer, this means that the knowledge that is in his or her body—often referred to by dancers as 'muscle memory'—is treated as a valuable asset. However, unlike memory, specific movement patterns become easier over time with repetition. As Lance 'Leftelep' Johnson described it: 'There are times when my body really aches, but it is actually easier to break than to walk to the grocery store because of how my muscle memory works.'[7]

One of the noticeable distinctions between the older and younger generations is the amount of time it takes for older dancers to warm up their bodies. This is a crucial factor to consider. For example, one b-boy reminisced about when he was in his early thirties and a younger b-boy in his early twenties called him out to battle. After six or seven rounds of back-and-forth dancing, the older dancer could see that the younger b-boy was running out of moves. However, the older dancer recalled that he was just getting warmed up and had a whole arsenal of moves that he was remembering as the competition continued. In other words, his vocabulary of movement was more extensive and as his body became increasingly limber through the act of competing, the younger b-boy was becoming exhausted, both physically and in terms of movement ideas.

In managing their ageing bodies, older dancers have to change how they interact, compete and prepare for competitions and events. Their newfound needs inform the practices within the culture as they explain their requirements to other participants, including some event organizers. This involves issues such as requiring more advanced notice before a showcase performance or competition to ensure that they are physically prepared and their bodies are warmed up. In this way, the older dancers also educate the younger dancers on their distinct needs, thus shaping the cultural practices they are passing down in new directions. This latter point is significant because it demonstrates that as b-boys and b-girls age they compare themselves to others on two axes. On one hand, they see that they have often remained in better shape than other people their age and that outsiders to this dance world perceive them as youthful. This results in positive body images, as reflected in their interview accounts, alongside occasional dismay at how 'youth' are treated, when outsiders treat them as possible troublemakers (a construct of youth) without recognizing their professionalism and correct age and life experiences. On the other hand, they experience challenges in their interactions with the younger dancers, whom they often teach. In this instance, they find ways to reorganize the structure or format at events or competitions in ways that make sense for their bodies, or they re-enact for youth how they used to 'battle' back in the day without proper warm-ups—something that resulted in injuries or pain.

In breaking practice, b-boys and b-girls do not regard one body type as the 'perfect body' or ideal shape. In fact, unlikely bodies are given 'respect' and 'props' for their skills, which are often body-type related. Taller b-boys and b-girls are given props for being able to do what traditionally has been conceived of as a shorter persons' dance form. Similarly, older dancers are respected for getting into the 'circle' (the dance floor). For example, Krazee Grandma is respected for starting the dance late but getting involved and participating in such an enthusiastic and dedicated way. Furthermore, some of the best b-boys and b-girls in the world would be categorized as having a 'disability' in their society, yet in the b-boy cypher or battle they are considered to be at an equal level with other top competitors. Each b-boy has a different body shape, size and level of ability—and some b-boys and b-girls dance and compete with missing limbs (sometimes using crutches) and degrees of deafness.

Goffman (1963) has argued that disability is created through the stigmatization of others. Here, one must make the distinction between constraints of the body as experienced through the ageing process, and constraints of physical or mental disadvantages that are overcome. In breaking practices, many of the most well-known dancers in the international scene have 'ill-abilities', a term defined by Luca 'Lazylegz' Patuelli as:

> The opposite of disability: adaptation of power, strength, and creativity. Anything one puts their mind to, will be done. Creating advantages from disadvantages: A physical or mental 'handicap' that one adapted themselves to, to pursue living a full, normal life.[8]

The international crew ILL-abilities was initiated by Luca 'Lazylegz' Patuelli. Lazylegz, who was recommended to me as an important figure in the Montreal b-boy scene because of his dance skills, event organization and professional development. Lazylegz has mixed his dance career with his work as a public motivational speaker about disabilities. The mission statement of ILL-abilities is:

1. To shatter any misconceptions that society may have about people with disabilities
2. To inspire and entertain audiences in a positive environment using unique motives and styles
3. To show the world that anything is possible; getting out the message, 'No Excuses, NO LIMITS'

The concept of ill-ability centres on a performance and engagement with the world of breaking competitions, showcases and choreography. Here, performances of athleticism are linked less with notions of 'youth', as has been discussed in the sociological literature (see Turner 2001), and more with notions of normality. In this way, accomplishments on the dance floor are set against the stigma faced by many people with physical limitations. One of the notable qualities of the b-girl/b-boy scene that I experienced was a lack of stigmatization in people's treatment of each other's body differences in the everyday training and performances that construct the local experience. Physical differences were not only 'accepted', but were set against a backdrop assumption that everyone is different from everyone else. This value system of acceptance is once again set against the backdrop of popular dance practices that are represented as competitive, misogynist and ego driven. However, breaking is characteristically open to varying forms of physical engagement, even with its high degrees of difficulty, and thus it is not surprising that ageing dancers have so easily found their place in the culture.

Retirement

The above discussion of disability highlighted the fact that physical constraints of the body are not necessarily deterrents to participation in dance culture, and not solely a feature of ageing. Rather, 'physical limitations' can also include injuries and illness, and ill-abilities are measured responses to the discussion of disabilities and dance. Another distinctive quality of the b-boy and b-girl scene addressed was that body limitations are set against a backdrop of inventiveness and an ethos of involvement. Ageing dancers must be understood within this context of openness and appreciation. The physical constraints of the body through ageing do not set limitations on a dancer's involvement in hip-hop culture. In fact, b-boys and b-girls often mentioned their frustration when outsiders asked them, in a way that seemed to be derogatory, whether they were still dancing. However, b-boys and b-girls recognized that by continuing their dance practice, their physical relationship—grounded in both the body and their identity—would change over time.

This section of the chapter discusses a further theme that emerged from the interviews and field research: the issue of 'retirement'. Three respondents discussed a change in their situation using this term. One b-boy at the upper end of the age category discussed his awareness that he would need to retire at some point. He continued by suggesting that he would give himself two more years of involvement. Two other b-boys described how they were done competing seriously, however, they would compete for 'fun' if their crew (i.e. their long-term friends) asked them to. One dancer responded:

> I have retired from my competitive battling career, although I would enter a crew battle if my crew asked me to. It's all about family and friends. It's just a hobby and pastime now. Except that I enjoy teaching master workshops and training people. I try to keep in shape and will do it as long as my body (and God) allows.

He continued to compete in local battles and perform in showcase battles internationally, and a year later he was invited to battle in a major international competition in France. What I would suggest from this self-reflection from the dancer, countered by the actions I observed, is that when it comes to breaking practice, retirement is more of a state of mind. In other words, he remained involved in dance performances yet saw himself as 'retired' from the stress involved in having aspirations to win competitions.

Thus retiring can be seen as distinct from quitting. Retirement involved a giving up of a particular sensibility towards competitive dance and improvement, with clearly articulated goals for development. Yet retiring did not mean giving up the dance. Those who retired would still continue to attend dance practices, judge events, teach workshops, compete and remain involved in dance in a variety of other ways.

Quitting, on the other hand, was regarded by those who continued as an impossible feat. As one dancer explained:

> Quitting breaking is a difficult thing to do. I've personally tried before, but couldn't find anything else to replace the wholeness that it instills in me. It is with you kinesthetically after so many years. My walk, sit, dress, everything is filtered through the dance. It would seem impossible to rid myself of that. Also, for many members of the crew it is the only way they know [of] accessing resources, such as attention from women.

Breaking's offer of various social functions (including access to the opposite sex), alongside sustained identity performances and understandings of the physical body, is mediated by shifting expectations within the dance practice. Older dancers are also expected to stay current in the trends within the wider context of hip hop culture worldwide.

The active choice to leave the dance seemed inconceivable, yet injuries were a constant threat if dancers continued to dance. In terms of ageing, one b-boy described being 'more prone to injuries, taking longer to heal, and longer to warm up' and another described:

> As a dancer the fear is of a serious injury that will prevent us working and in the worst-case scenario end our career. B-boying is an abrasive dance form which takes its toll on the body. I have to be realistic about what age I can continue dancing until, and try to do everything I can to extend that window.

Some of the strategies involved in extending the period of life within which one can continue to dance include an acquisition of knowledge about care for the body to maintain and improve performances. As Robby Graham of Badtastecru explained:

> Physically I still feel really good. As I have gotten older I have learned more about nutrition, stretching, and how to look after myself . . . Working in a professional context, having access to information from colleagues, etc., has been invaluable in learning that lesson.

Likewise, at a recent panel discussion in Toronto,[9] Mariano Abarca's advice to younger up-and-coming dancers was to 'stretch, stretch, stretch'.

A dancer described the advantages and disadvantages inherent in the physical changes of ageing in terms of various types of effort:

> As far as physical restraints, my body has adapted to breaking so much so that it is literally instilled in me. Therefore, it takes less physical effort for me to conduct a flow than it did when I was younger. New power moves can be tricky, and harmful, to acquire at my age, but my body has the experience of the dance in it. Older breakers can usually point [out] who has been dancing for a while by the way they move in their own way and how they interpret the music. Therefore, experience can balance physical limitations I feel from getting older.

On the one hand, there are perceived constraints on what is physically possible. On the other, there is an acknowledgement that knowledge accrued through participation provides new pleasures in experiencing the dance as a spectator.

Top performers may peak physically in their early twenties. However, their social status often operates in such a way that their reputation does not peak until their early thirties. These variations challenge socially and scientifically prescribed notions of 'peak age', and they operate through the dancer continuing to practise and improve skills in particular domains later on in their careers. This partly has to do with the work of building an international reputation—work that involves factors other than the physical body and performance, such as self-promotion and peer confirmation.

Most of the older top performers I interviewed mentioned having early success in their practice. As individuals or with their crew, they often had won the first amateur battle they entered. Now, younger generations of dancers have a very different experience with the dance in an ageing culture. The likelihood of winning a competition would depend on a situation where only amateur competitors could enter.

Ageing and Musical Meaning

Informal education in popular dance practices tends to involve some degree of knowledge acquisition based around musical competence. Older b-boys and b-girls mentor younger dancers not only on techniques and aesthetic qualities, but also on what is 'good' music. Part of what is considered to be a 'good' education in b-boying/b-girling practice relates to understanding and appreciating—indeed 'listening' to—the music as much as it does to gaining competence in dance.

In interviews with older dancers—those now in their forties—participants acknowledged that when they first began breaking, their practice was done during the party at the moment when the DJ would loop a break beat (the percussive break in a song where parts of the rhythm section, or less frequently the horn section, 'go off'). B-boys and b-girls would match this moment in the music by 'going off' themselves. This is how breaking emerged originally as a practice in New York City. As the story goes, the DJs began to extend the breaks and b-boys and b-girls would stay down on the ground longer doing moves.

However, as dancers got older, they started to think about the meaning of the songs differently, and they did this by making sense of the lyrics and truly listening to them. Dancers also began to put the meanings of their favourite songs into a historical and cultural context.

For example, take the classic b-boy break, Herman Kelly & Life's 1978 song 'Dance to the Drummer's Beat'. One b-boy from New York City explained that as the scene started to age, and he felt that people were not dancing anymore—or dancing to the breaks of records—he noticed what the lyrics of the full song actually were—'dance to the drummer's beat'—and he saw these lyrics as meaningful. This is what the b-boys and b-girls did, literally, and it was right there in the song the whole time (he just had not noticed the words). He used this song as an example, when teaching younger b-boys and b-girls, to foreground breaking not to the full song but rather to the break of the song that is looped by the DJ. This was an educational experience for other DJs, who did not know how to play music for b-boy events. The music 'is the key' to cultural memory. Older dancers want to remember not only the pleasures of youth but also the continuing pleasures involved in thinking about dance practices in new ways with new people.

With these scene mutations, something strange occurs in b-boying scenes worldwide: the musical tastes of the group resist ageing. Instead, musical tracks from the 1970s and early 1980s remain perpetually present and popular. Also, the longer ageing dancers dig into the popular music practices of their youth, the more significant the lyrics become to their musical understanding. This is unusual for a cultural practice of dancing to break beats. What is special about that music from the early 1970s is its inclusiveness. The beat continues, but the lyrics and full orchestral backing in the fills refer back to earlier forms of popular music. Similarly, older b-boys and b-girls inform the musical tastes that dominate the contemporary scene.

Conclusion

Simon Frith (1987) once wrote that if you never got into popular music you were never really young. The logical extension of this might be that if you never 'grow out of' popular music tastes (and continue to maintain your dance practice) then it is like drinking from the fountain of youth. However, the body does wear. And like the quest for the fountain of youth, I often came across dancers who had the desire to become young again—that is, to get back into their dance practice, often after time away from the dance and the dance scene. At the same time, all the dancers I spoke with who were aged over 40 agreed that the dance had progressed so rapidly that they could no longer keep up on a technical level. Those dancers who never stopped, or took breaks from the dance (for reasons such as injuries, busy careers in other fields and/or raising children) only to return again, could not imagine their lives without this dance at the centre of their identity and relationship to music.

I suggest that the various forms of learning about the dance, whether self-taught, peer-to-peer or mentored, have all been aspects of this popular dance practice since its inception. The category of 'youth' provided by earlier subcultural theories has downplayed the importance that older teachers have always had in the development of new forms of dance. The technical training and the hours invested by dancers in practice with others in their 'extended family' make aesthetic expression possible. This aesthetic expression involves a musical competence involving matters of taste, identity and collective judgement.

With the continuation of this form of dance, the familial analogy has changed from a one-dimensional signification of the closeness of friends who participate in a shared activity to what I argue are extended families that are multigenerational in composition. In other words, hip hop is old and is not looking too bad for its age. The idea of the family has also extended to incorporate the 'family' of hip-hop and funk styles including popping, locking and, in the contexts of particular competitions and club nights, other styles of dance such as house, new style, 'hip hop' and krumping.

As Hennion (2007) observes, concentration does not need to be treated as a merely psychological factor. Part of self-reflexivity involves the body being revealed to itself through gestures and performance. For b-boys and b-girls, taste is a self-reflexive performance revealed through activities that develop technical mastery, which lead to aesthetic expression. This taste performance is then tied to the social group that shares aesthetic preferences; in combination, these aspects constitute a category of 'youth' involvement for those who sustain their practice. Instead, these tastes are shared in the self-reflexive construction of a crew politic that prevents dancers who have 'retired' from ever truly 'quitting'. In the cyphers, concentration is a shared experience involving like-minded individuals with similar musical tastes.

Ageing and retired dancers make up a large body of the consistent spectators for breaking events. This is crucial because these audiences, by having acquired knowledge about the dance, support the dance and strengthen the case for breaking having its own evaluative frameworks. Aesthetics get performed in both cases. Dance is a performance of musical tastes, and older dancers perform judgement in a way that younger dancers cannot—not just judgement in terms of judging competitions but in terms of what they are still doing and teaching, and the music to which they still respond, and how they respond. Their age and experience add 'depth' to peer judgement, as well as to the dance itself, since their 'aged' dancing is still defining an aspect of the form. And although they cannot physically move in the ways that they once could, their bodies remember and they know that one of the most important criteria for aesthetic judgement is not looking at dance, but listening to it—and listening, I would argue here, is probably the quintessential 'embodied' experience, for young and old alike.

–5–

Slamdancing, Ageing and Belonging

Bill Tsitsos

Introduction

Slamdancing and moshing are forms of dance associated with the punk and hardcore music scenes. In both slamdancing and moshing, 'participants (mostly men) violently hurl their bodies at one another in an area known as a "pit"' (Tsitsos 1999: 397), close to the stage where bands perform. Although the two terms are often used interchangeably, slamdancing and moshing are, in fact, distinct dance styles. On one hand, slamdancing tends to be a more frenetic style of dance. Dancers sometimes run in a counterclockwise circle (a 'circle pit') while occasionally colliding with each other. At other times, 'slamdancers do not run in a circle, but rather move in a more "run-and-collide" fashion, simply throwing themselves into the part of the pit where the most people are gathered' (406). On the other hand, moshing is not as fast paced as slamdancing. The body movements involved in moshing tend to be more exaggerated, individualistic displays, sometimes involving jumping kicks that seem to be derived from martial arts (406). Despite these distinctions, however, both slamdancing and moshing are characterized by their overwhelmingly male and youthful participants.

To date, no one has explored the significance of slamdancing or moshing for older scene members. Drawing on empirical data collected during interviews with older (defined here as age 25 and above) members of the hardcore punk music scene, this chapter presents the first analysis of the relationship between slamdancing and ageing. The chapter begins by discussing one scholarly work (Fonarow 1997) that examined audience members' shifting choices of where to position themselves at music gigs as they age. Although Fonarow's work is clearly relevant to this project, I expand upon her exclusive focus on audience members to show how older people remain active in the scene by playing in bands, booking shows and sometimes slamdancing. The second part of the chapter looks directly at why (and when) some scene members do slamdance. I liken the dancing to an experience of Durkheimian 'collective effervescence', which reconnects older scene members to their friends, who are often on stage performing, as well as to an idealized memory of the scene from their youth. Older scene members describe their feelings of being out of place in the scene and relate these feelings to the dancing experience. The chapter then takes a closer look at the ways in which some interviewees perceive that slamdancing has changed (for the worse) since the time when they were younger. This widely expressed view reveals how the interviewees' relationships to the scene have changed as they themselves have grown older.

During the summer of 2010, I interviewed twelve individuals (eleven male, one female), between the ages of 26 and 38, about slamdancing. My initial plan was to define 'older' scene members as age 30 and above. However, over the course of informal conversations with various acquaintances about ageing and the punk scene, I found that people over the age of 25 often expressed a clear sense of being on the margins of a predominately youth culture. So I decided to broaden my interview criteria to include individuals aged 25 and above. Table 5.1 shows the name (no real names are used) and age of each respondent, as well as other information to which I will refer later. Seven of the interviews (1–7 in Table 5.1) took place in person, and the remaining five were conducted by email. Although I used the term 'slamdancing' in my interview questions, some interviewees used the term 'moshing', indicating that the dances are often regarded interchangeably. Only one interviewee (the youngest respondent) still slamdanced regularly, and four respondents admitted to dancing either 'very rarely' or 'sometimes'. However, many respondents—including some who answered 'no' when asked whether they still danced—told stories of returning to the pit at some point since forsaking the dancing. I asked them to elaborate upon these stories in order to look for any patterns that emerged.

Table 5.1. Interviewee information

Name	*Age*	*Still dance?*	*Play in a band now?*
1. Johnny	28	No	Yes
2. Ian	34	Very rarely	Yes (and operates two record labels)
3. Ray	31	No	Yes
4. Keith	26	Yes	No
5. Dez	27	Very rarely	Yes
6. Henry	31	No (never did)	Yes (and operates a record label/store)
7. Joey	33	No	Yes (and books shows as a job)
8. Ron	33	No	Yes (and operates a record label)
9. Shawn	27	Very rarely	No
10. Glenn	30	No	No
11. Tina	32	No (never did)	Yes (and operates a record label)
12. Milo	38	Sometimes	Yes (and operates a record store)

From these interviews, I discovered that returning to the pit is a way for older scene members to reconnect with friends (some of whom are performing onstage) and with the scene itself. In this sense, slamdancing assumes the power of Durkheimian 'collective effervescence' in creating and reinforcing collective solidarity. Although Durkheim presented the concept of collective effervescence as part of his sociological analysis of religion, it is also useful for understanding other, nonreligious phenomena. For Durkheim, the key function of religion is the formation and reinforcement of social solidarity. In fact, he asserted that when people worship a god, they are in fact celebrating the social group to which they belong, with 'the god being only a figurative representation of the society' (1995: 227). So Durkheim's arguments about religion can be applied to the study of social solidarity more broadly. Times of collective effervescence are characterized by heightened group excitement, when individual identities become less important than group affiliation, and it is during these times that social solidarity is strengthened. My interviewees' accounts of returning to the pit revealed the ways in which dancing serves to reconnect them to the scene in such a Durkheimian fashion.

In their explanations for why they no longer slamdance as often as they once did, my interviewees also detailed how active members of the scene view their changing roles in the scene as they get older. In discussing dancing, interviewees cited their disillusionment with perceived negative changes in the scene (reflected in increased violence in the pit), as well as their concerns over age-related physical limitations. In other words, it is not just they who are changing by growing older; the scene is supposedly also different. One potential reaction to this would be to distance themselves from the scene, but as many of my respondents were actively involved in the scene by playing in bands and so on, this often was not an easy option. Returning to the pit as an older person provides a way for older scene members to symbolically reconnect with the scene. This is especially true when the band that is performing consists of their friends, is a reunited band that was popular from their youth or is covering a song that they enjoyed when they were younger. However, the scene to which scene members are reconnecting in the pit is as much an idealized version of the scene from their youth as it is today's scene.

Space, Ageing and the Punk Music Venue

On one hand, audience members' decisions of where to stand and whether or not to dance at a show reflect their feelings about the performer. At the same time, these decisions symbolize audience members' sense of belonging in the larger music scene. In 'The Spatial Organization of the Indie Music Gig' (1997), Fonarow argues, 'At the indie gig, where one places oneself is a physical enactment of a *statement* of assessment and allegiance' (1997: 368) directed towards the performer. However, Fonarow also shows how age is an influential factor behind these placement decisions. She divides the audience area into three 'zones', each with its own age profile. Most dancing takes place in zone one, which 'is the domain of the greatest and most frenetic activity, the youngest audience and strongest statement of fanship' (1997: 361). In Fonarow's venue diagram, zone one includes the area directly in front of the performer, a 'pit' area behind that and a centrally located 'mosh pit' at the core of the pit (1997: 363). Choosing to watch performers from within this area 'is a public assertion of alignment to the band' (1997: 364).

Behind zone one is zone two, which is characterized by an older crowd than zone one. The people in zone two do not necessarily feel less enthusiastic about the performer; rather, most people in zone two 'privilege their spectatorship in terms of aural connoisseurship and the undivided attention given the performers' (1997: 366). In other words, the people in zone two praise this particular zone for offering the best sound quality and the ability to watch performers without being jostled by dancers. Whereas people in zone one value the 'tactile, aural, and visual' (1997: 366) experience of live music, the tactile dimension becomes less important for audience members as they age. Nevertheless, according to Fonarow, even the people in zone two view their choice of location as superior, because 'each zone uses the same idiom of closeness to the band to privilege its position in the participant framework' (1997: 368).

Zone two is in front of zone three, which 'reaches into the back of the venue, where numerous activities that do not directly pertain to the performance take place' (1997: 366). The fact that the bar is located in zone three in many clubs is a major attraction of this zone for people who are old enough to drink legally. The people in zone three include various music 'professionals' (1997: 367), such as booking agents, promoters and journalists, as well as 'the oldest fans (aged in their late twenties to thirties), individuals who are engaged in different activities other than watching the band, those individuals who dislike the performance' (1997: 367) and others. Even though zone three is the most physically distant from the performers, people in this area (with the exception of those who are there because they do not like the performers) are similar to the people in the other two zones in that they too privilege their location in terms of some form of closeness to the performers. The closeness for people in zone one is physical closeness, while the closeness in zone two is undivided attention. In zone three, especially among professionals, the closeness takes the form of 'stage passes that give them access to the after party or to the back stage' (1997: 367).

Overall, the depiction of ageing in Fonarow's essay is of a process which 'is seen to marginalize one's ability to participate' (1997: 365). The trend over time is for audience members who spent their earlier shows positioned close to performers, in zone one, to move progressively backwards as they age, 'until they are aged out of the venue all together' (1997: 369). Fonarow points out that 'it is very rare for older individuals to be seen in the front' (1997: 365), and her conclusion is that 'gigs are an age-set activity not separable from the general culture, but nevertheless a fundamental ritual event in the constitution of youth as a social category' (1997: 369). For her, the gigs are centred on youth participation, and older people are way up the back, if they are present at all.

Analyses of ageing in the independent/punk music scene, such as Fonarow's, liken the ageing process to moving further towards the back of a venue before finally outgrowing the entire live indie music experience. This ignores the role of musical performer, thus failing to take into consideration the fact that the performers of the music around which the scene is centred are often themselves in their late twenties and thirties. Rather than moving further towards the exit, these people are literally front and centre at shows. Indeed, in my search for individuals to interview for this project, I could find very few people over age 25 who still attended shows but were not also involved in the scene as a band member, a record label operator, a show booker or some similar occupation. As the last column in Table 5.1 illustrates, nine of my twelve interviewees were members of bands when I interviewed them. Of those nine, six were also involved in the scene on what might be termed the 'production' side, by booking shows, operating record stores and/or operating record labels. When I asked my interviewees whether they knew of anyone over the age of 25 who still attended shows but did not play in a band or do any other work in the production area, only one interviewee could think of someone. Unfortunately, that individual did not agree to be interviewed for this project, but there clearly are some people who remain tied to the scene solely in the role of 'fan'. Nevertheless, it is also clear that there is a concentration of older scene members performing in bands, booking shows and so on. It is possible that the punk scene's Do It Yourself (DIY) ethos contributes to this by legitimizing older people's continued presence in the scene, as long as they are still 'doing' something. Some of these older 'doers' know each other and, importantly, they sometimes return to the pit when each other's bands play. Perhaps those individuals who are no longer productive members of the scene, doing work that helps sustain the scene itself, more easily find themselves marginalized and aged out of the back of the venue.

Returning to the Pit

Fonarow (1997) claims that an important part of the 'constitution of youth' process is related to being in zone one at shows. What, then, are we to make of the times—as rare as they may be—when older audience members choose to venture into zone one? Fonarow classifies the performers as 'ritual objects' (368) that are being exalted by the crowd in zone one, but in my interviews, older scene members described returning to the pit as a form of reconnection with friends. It is not just the performers who are being celebrated, but rather the shared experience with others in the crowd. Recounting his experience at a Jesus Lizard reunion show in 2010, Johnny (age 28) stated, 'Everyone was pretty into it . . . so if you were in the crowd trying to get to the front of the stage, you were basically slamdancing, not even necessarily on purpose.' Ray (age 31) described his last time in a pit, when a band covered a song by the well-known

1990s Swedish hardcore band Refused. According to Ray, 'I ran around and sang along.' When asked to elaborate on the difference between that and dancing, Ray revised his description of what had happened, saying instead, 'It was more like I was participating in the singing along.' Meanwhile, Dez (age 27) last entered the pit at a show by the Canadian band Propagandhi in 2009. He summarized the pit as 'wild without being overwhelming or violent'. When I asked him what kind of motion(s) he was making, he said that he was 'sometimes jumping on people to sing along, running back and forth, grabbing your nearest friend and just kind of shaking them, riling them up, screaming the words in their face'.

The above descriptions of venturing back into the pit and reconnecting with the scene call to mind the concept of 'collective effervescence' outlined by Durkheim in *The Elementary Forms of Religious Life* (1995). During periods of collective effervescence, when 'individuals are gathered together, a sort of electricity is generated from their closeness and quickly launches them into an extraordinary height of exaltation' (1995: 217). For Durkheim, the strengthening and celebration of community are the core functions of religion (1995: 44), and, 'It is in these effervescent social milieux, and indeed from that very effervescence, that the religious idea seems to have been born' (1995: 220). Individual identities are subsumed under group membership at these times, whether they take place in the pews at a church service or in the pit at a punk rock show. Durkheim conveys a sense of this power of communal motion in discussing the phenomenon of collective effervescence when he writes that 'a collective emotion cannot be expressed collectively without some order that permits harmony and unison of movement, these gestures and cries tend to fall into rhythm and regularity, and from there into songs and dances' (1995: 218). Even though he claimed to no longer slamdance, Ray admitted that the dancing 'can actually be kind of beautiful, actually, because we [the band] are playing and going through some kind of catharsis, and they [the crowd] are dancing and going through some kind of catharsis. So it's an extension of the music.' Ray's words are another indication of the Durkheimian connection to something larger than the individual that dancing can evoke and encourage at a show.

Accounts such as Ray's elevate the live music performance to the status of religious ritual. Fonarow (1997) seemingly would agree with this argument—for instance, when she asserts that performers are 'ritual objects'. However, based upon my interviews I take a more Durkheimian approach by arguing that, at least for scene members who return to the pit later in life, the dancing does not just exalt the performers. Just as Durkheim indicates that the worship of god is, in fact, the worship of the community, my suggestion here is that the return to zone one is as much about celebrating and reconnecting to the scene itself as it is about exalting particular performers. This was certainly the case for Ian (age 34), who described his last time slamdancing, which was at a reunion show by a band that had been popular in the 1990s and early 2000s. Ian went to high school with the band's singer and booked multiple shows for the band when it was actively touring. After they broke up, the band played a few reunion shows, but Ian never attended any of them, until his own band also played at one of the more recent reunions. At the show, when his friend's band started playing, Ian was sitting at his own band's merchandise table with his wife, towards the back of the venue. When Ian's wife asked him whether he was going to go up front to watch the band, Ian described his response as, 'I don't know. I'm too old for that. I'm happy just sitting back here and watching.'

About halfway through their set, however, he changed his mind. In his words, 'They were just playing like hit after hit after hit. I was having a blast so I moved up front and just started jumping on people and climbing around, jumping off stage, and acting like it was ten years ago. I had a blast.' Ian had by no means become disconnected from the scene in the ten years leading up to our interview. He had steadily performed in bands, and he operated two record labels at the time we spoke. Nevertheless, Ian described finding it difficult to relate to the younger audience for his current band. The reconnection to the scene that the dancing seemed to provide for Ian was a literal reconnection with other audience members, who were mixed in age, although 'the majority of the people were young'. I argue that the process of physical reconnection with dancers in the pit constitutes a symbolic reconnection to the scene itself. This duality mirrors religious ritual as it is described by Durkheim (1995), which is about worshipping physical sacred objects—such as the band, in Fonarow's (1997) view—as a means of celebrating the community.

To the extent that older individuals are, in fact, reconnecting to the scene when they dance, it is important to keep in mind that the scene with which they are re-connecting is often an idealized vision of the past. For instance, the stories of many of my interviewees suggest a general pattern for when older people decide to return to the pit. The most recent dancing experiences of individuals such as Johnny, Dez, Keith, Ray and Glenn took place either during performances by bands that were popular when my respondents were younger (either teenagers or in their twenties) or during performances of cover songs that originally were recorded by bands when my respondents were younger. Johnny danced during a reunion show by the Jesus Lizard, one of the biggest independent bands from the early 1990s. Dez danced to the Canadian band Propagandhi, a band he has enjoyed 'since forever' but which has rarely played in the United States since its first recordings from the early to mid-1990s. Describing his affection for Propagandhi, Dez stated, 'My interest in them has not waned at all. It's just been so steady. It just seems like the older I get, they're obviously getting older, and the music's just getting darker, and the lyrics are getting angrier, more fine-tuned and directed.' For Johnny and Dez, the lure of these bands they have enjoyed for so long was enough to draw them back into the pit.

In an email, Glenn recounted the story of the last time that he danced, during a reunion performance by the 1990s California band Unbroken at one of the shows to celebrate the release of the book *Burning Fight*, about the 1990s hardcore scene. The prospect of an Unbroken reunion was particularly surprising and exciting to many fans, because one of the band's guitarists had committed suicide in the late 1990s:

> The last time I was 'in the pit' in an intentional way was during Unbroken's reunion set at the Burning Fight fest in Chicago. I think it was a little bit of nostalgia, seeing a band that was so incredibly important to me when I was younger and being reminded of the feeling that brought me into hardcore in the first place. Besides, seeing all of the other 'old folks' of the scene stage diving and moshing was both exciting and a little touching. I jumped off the stage during 'In the Name of Progression' and moshed my way through the crowd and back on stage. It was a brief thing, but it was a blast and it brought me back to that feeling of abandon and joy that I used to get when I first started going to shows.

Perhaps the presence of other older folks in the pit during performances by older bands emboldens some people to re-enter zone one, but none of my other respondents mentioned that. Instead, they focused on the band that was playing at the time. Ian, for example, danced for a band whose members included friends from his youth. In the stories discussed above, Johnny, Dez and Glenn all danced to bands with which they felt a connection from their younger years. Ray and Keith, meanwhile, did not necessarily feel as strongly about the bands that were playing when they chose to dance again. Rather, they felt a stronger attraction to the songs those bands were playing; as discussed below, these were covers of songs by bands from the interviewees' younger years. Specific bands and/or songs have the power to draw older scene members into the pit by invoking a particular memory or feeling of what the scene was like when they were younger.

For example, Ray felt compelled to dance when a newer band covered a song by the early to mid-1990s Swedish band Refused, which was one of the most well-known European straight edge (rejecting drugs and alcohol) bands of the time. This was enough to draw Ray, who is also straight edge and a fan of Refused, into the pit. Meanwhile, Keith's most recent pit experience also occurred while he was watching a band cover a song, 'Rise and Fall', by its singer's band from the 1980s, Leeway. Keith was 26 years old at the time of our interview, which means that he was an infant when Leeway formed in 1984. Nevertheless, he is enough of a fan of that era of hardcore music that his excitement over the song was palpable as he described it to me during our interview. In our interview, Keith's responses indicated that he still dances more often than my other respondents. Perhaps that reflects his status as my youngest interviewee. As a (relative) youngster, he is less 'out of place' in the pit, and less susceptible to physical injury—two issues that I will discuss below.

Physical Limitations and Being Out of Place

One topic that I expected to hear more about during interviews was that of age-related physical limitations, which could affect respondents' ability to slamdance. Only three of my interviewees referred to the physical consequences of getting old. Among the interviewees who cited their own physical limitations as a reason for less (or zero) participation in the pit, Shawn (age 27) mentioned his poor physical fitness, which revealed itself the last time that he was in the pit. His most recent dancing experience was at the last show by the band Have Heart, which formed in the early 2000s, and Bane, a band that has existed since the mid-1990s. Shawn wrote to me via email:

> The last time I danced was in October 2009 at the last Have Heart show. I only danced for Bane and Have Heart. The best way I can describe what happened was many muscle cramps, shortness of breath and not being able to pull myself up to crowd surf (though I contribute that part to there [*sic*] being so many people just pushing forward). I was just so out of shape from not doing that type of activity often enough. By the end of the night I looked and felt like I had just sprinted an entire marathon. I had a blast though. Before that, I can't even remember the last time before that.

As a side note, Shawn's story does fit the profile, described above, of respondents' returning to zone one for bands that hold some personal significance from their youth. Even though Have Heart was not a particularly 'old' band, Shawn (age 27) was relatively young among my interviewees. He was 19 when Have Heart formed, so for him Have Heart had enough nostalgia appeal to make all of the physical pain worthwhile. I asked Dez, another one of my younger interviewees at age 27, about the issue of physical limitations, and he stated, 'I have gotten hurt before, since I've been older. I used to catch a punch in the mouth and you know, get a bloody lip and be like, "OK, hold on. Give me like 2 minutes." And then I'd be right back and I'd be like, "Hell, yeah!" . . . I'd be pissed off if that happened now.' When I asked Joey, who plays in a band and has a job booking shows, whether he still slamdanced, he answered:

> Absolutely, positively not. I still think it's fucking cool. I do shows basically for a living . . . I still think it's fucking cool to see a room full of like young kids, boys or girls or whatever, going fucking haywire to a band. I might think the band sucks, but I also understand that, let's not kid ourselves, punk is a young kids' game. It doesn't mean that you stop being punk, but some people are like, 'Punk changed'. And it's like, 'Yeah, it did, and you either quit and you got out of it.' It didn't stop. Punk keeps going. You might not like what it is now, but it keeps happening. There's always going to be young kids getting into it . . . I like to see that. Would I do it myself? Not really. Not anymore. Now I'm too worried about breaking my knee or breaking an arm or something.

Even though he still contributes to the scene by playing in a band and booking shows as his job, Joey said he believed ageing marginalized people from the scene. Similarly, when asked whether he enjoyed many newer hardcore bands, Ian stated, 'I get stoked on new bands, but . . . I wouldn't be up front going crazy for them', in part because 'everyone else is about half my age'. Among interviewees such as Joey and Ian there exists a sense that the pit is not the appropriate place for older people.

This feeling about the dancing reflected a larger sense that the scene had changed. Ian described this by stating:

> When I got involved, punk and hardcore separated themselves as having more than music, social importance, whether it was anti-racism, animal rights. It's pretty rare to see that these days.

As a result, he described sometimes finding it difficult to relate to the audience for his own band. Nevertheless, despite such feelings of disconnection from younger scene members, there remained a strong sense of connection to the scene itself as an institution among my interviewees, many of whom did work that was necessary to keeping the scene alive and functioning. Other interviewees also conveyed stories of their own conflicted feelings regarding their relationship to the scene and slamdancing. Specifically, as discussed below, they voiced criticisms of the behaviour of other dancers in the pit and contemporary pit etiquette (or the lack of it).

Criticisms of Violent Dancing

Some interviewees described how they came to feel at some point that the scene had changed for the worse. Joey, in discussing changes in the scene around the year 2000, lamented the 'weird thug mentality that took over hardcore at that point, and that really alienated me'. This change was reflected on the dance floor by an increase in violence and a lack of concern and respect for others at shows. Dez, Glenn and Johnny also expressed a belief that dancing had grown more violent since their first experiences in the pit.

Ray, meanwhile, identified the turning point when he decided to stop slamdancing at a show where there was a 'homogenous group of white males dominating the room'. Even though the boys were already occupying '75 per cent' of the floor space, they still threw their bodies at the people who had pushed their bodies against walls of the venue to watch the bands play. Among my respondents, the stories of disenchantment with the scene conveyed a perception that the community orientation of the scene has declined. Dez raised the issue of the Internet as a forum for a type of community that actually could contribute to a weaker sense of solidarity with those people who are geographically closest. Glenn and Ian explicitly cited violent 'kickboxing' in the pit as a factor behind their respective decisions to stop dancing. Glenn's description of his earliest slamdancing experiences conveys a sense of just how important the dancing was in making him feel a part of the scene community:

> I was always a nerdy non-athletic kid and punk shows, along with skateboarding, had this physicality that I wasn't accustomed to but that I really liked. It felt somehow safe, too—despite all of the flailing arms and legs and the days of sore muscles and bruising after, I was never worried about actually getting hurt. For those first few shows, it was

> pretty much entirely about slamdancing—the music was just an excuse to run around in a circle and throw myself into people.

For interviewees such as Dez and Glenn, disillusionment with the dancing paralleled growing dissatisfaction with the scene in general. Admittedly, it is common for people to lament the ways in which their scene, whether it is punk rock or another subculture, has changed for the worse since the 'golden age' of their youth. Upon closer inspection, it is possible to find evidence that the scene was not always so communally oriented in the past. For example, in explaining why he stopped slamdancing at age 17, after having been an avid dancer since age 13, Joey described seeing Bikini Kill in concert as a turning point. The band's aggressive stance against male dominance of the scene, including the pit, inspired Joey to view a subsequent live show from a different perspective:

> I stepped back and, like, watched what was going on, and then I vividly remember being at a show . . . there was fight while they were playing, and I remember it was a fight over kids dancing, and the band didn't stop, and kids were just clobbering each other . . . I noticed a definite divide, and that was like a definitive moment, seeing that, where I genuinely was just like, 'I want to be part of this [scene], but I don't want to be part of that [dancing].' To be totally honest with you, I probably started to turn into an elitist dick.

Joey was 33 years old when I interviewed him, so the Bikini Kill show that he described had to have taken place around 1994, suggesting that there has always been a violent, individualistic element to the scene that can manifest itself in the pit. For example, Dez criticized the excessively violent dancing of kids today during our interview (see the quotation above) while also recalling receiving 'a punch in the mouth' in the pit when he was young. Back in the 1990s, there were probably people wishing for the good old days of the 1980s, when people supposedly danced less violently. The key point to take from these criticisms, then, is not necessarily that dancing and/or the scene have changed, but rather that the individuals voicing those criticisms have themselves changed in terms of their perspective of the scene. Nevertheless, despite these changes, my interviewees still maintained some connection to the scene—often as performers in (and fans of) bands—and the pit remained a place to occasionally renew that connection for many, despite their misgivings about the dancing.

It should be noted that the critical position towards slamdancing is not universal among older scene members, including my respondents. There are people who return to the pit for something other than a positive feeling of connection with others. For example, Milo (age 38) recounted three of his most recent stories of dancing (all from within the last five years), and two of them involved older bands. At one show, the band Slumlords covered a song, 'Sick People', by its singer's 1980s band, Breakdown. One of Milo's other pit stories took place at a show by the band Integrity, which has been active intermittently since the late 1980s. The song to which Milo danced, 'March of the Damned', originally appeared on an Integrity album from the early 1990s, so Milo's decision to dance to that particular song lends support to the notion that older scene members return to the pit as an act of nostalgia and reconnection with the scene of their youth. Milo described these two experiences via email in response to the question, 'When do you dance?'

> Rarely, but sometimes the right mix of band and song and crowd brings it on. Seeing Slumlords play 'Sick People' by Breakdown at their last show caused me to try to punch every person near me, dance hard, and when I came to I was holding the stage monitor [a large speaker] over my head like I was going to smash it on someone. Luckily I came to my senses and just put it back down. Seeing Integrity play 'March of the Damned' also caused me to try to hit everyone around me and go completely mental.

For Milo, these show experiences seemed to generate an altered state of consciousness defined by tremendous hostility. Perhaps this had something to do with the bands—both of which could be classified as examples of the 'weird thug mentality' in hardcore, to quote Joey. This, however, does not fully capture the variety of Milo's dancing experiences. He also wrote that 'seeing the pop punk band the Ergs play always caused me to happily and nonviolently stage dive like crazy for every song'. Milo is certainly not the only older scene member who sometimes returns to the pit looking for something other than a positive communal experience. He was, however, the only one of my respondents to admit to this, suggesting that he is rare among older scene members.

Conclusion

For most of my interviewees, the return to the pit is a return to an idealized version of the scene as many of them remembered it from when they were younger. Recall interviewees such as Ian describing himself as 'acting like it was ten years ago' while dancing. Similarly, Glenn wrote that dancing while Unbroken played 'brought me back to that feeling of abandon and joy that I used to get when I first started going to shows'. For many of these older people, however, the scene is not just a part of their past, but remains very much an important part of their present identities. As one interviewee, Ron (age 33), wrote by email, 'Almost all of my friends that I've kept throughout life came to me by way of our shared love of underground music. It's how I spent every day from age 13 on.' Ron was not the only interviewee to share such a sentiment, and for many respondents, re-entering the pit is a way to overcome conflicted feelings about perceived changes in the scene (and about 'age-appropriate' behaviour) and to celebrate connection and collectivity. Insofar as it reconnects participants to something larger than their individual selves, specifically the punk/hardcore scene, slamdancing is an example of Durkheimian collective effervescence. For older scene members, especially those who do the less 'effervescent' work to sustain the scene (such as operating record labels and stores), the energy of the pit can provide a reminder of why that work is worthwhile.

Aaron Cometbus is a writer and drummer who has been performing in various punk bands and writing his eponymous fanzine since the 1980s. In *Cometbus* issue no. 54 (2011), he described the experience of accompanying the hugely successful band Green Day as a guest on an Asian tour. Cometbus was a mentor to the members of Green Day in the 1980s and had served as a roadie on one of their earliest tours of the United States, before they achieved stardom. In the years after the band became popular, Cometbus and the band grew apart, but they invited him on the Asian tour nonetheless. Cometbus described the experience of slamdancing together with the band members in a bar in Japan to various 1980s punk records:

> Eighties punk was our native soil, out mother's milk, regardless of the different ways we related to it and the different directions our lives had taken in the decades since. What we still had in common was greater than our differences, or so it seemed in Osaka as we flew through the air or found ourselves flat on our faces on the floor. We were like landsmen from a country that no longer existed, performing all the old rituals and finding that they still made sense and still spoke meaningfully to our needs (2011: 81).

In this passage, the power of the dancing to reconnect older scene members to their friends is clear. In fact, the dancing ended with Aaron and Billie Joe (Green Day's singer/guitarist) exchanging 'a long and tender kiss' (2011: 82) in the middle of the impromptu dance floor. Beyond just reconnecting with his friends, however, the dancing revived his sense of connection to the 1980s punk scene, his 'mother's milk'. My interviewees expressed similar feelings about returning to the pit and re-newing their bonds with the scene that meant so much to them in their formative years. Slamdancing is the physical enactment of this connection, which makes the emotional bond possible.

–6–

Rock Fans' Experiences of the Ageing Body: Becoming More 'Civilized'

Lucy Gibson

Introduction

Rock music has been a key feature of popular culture for over fifty years. Moreover, despite rock's endurance, it has frequently been portrayed as youth music and as symptomatic of teenage angst (e.g. Frith 1983; Bradley 1992; Keightley 2001). This chapter aims to move beyond such representations by exploring some of the ways in which ageing, with particular reference to embodiment and corporeality, shapes and impacts on older fans' participation in rock music scenes. The discussion offers evidence of how those in a sample of self-defined rock fans negotiate certain physical constraints and adapts behaviour as they grow older. Bradley (1992: 119) suggests that 'rock music and youth culture went together because rock music was the most central constituent of youth culture and force for intensifying it and spreading it'. Over twenty years ago, Frith (1983: 9) suggested that the 'sociology of rock is inseparable from the sociology of youth'. I am not attempting to suggest that this distinctive relationship between youth and rock music is incorrect. Clearly, this familiar contention is historically accurate. Nonetheless, I conducted ethnographic research into 'older' rock music fans in the United Kingdom (during the period 2006–09), which illustrates that many rock fans are no longer 'youths', and ultimately challenges this longstanding assertion.

Within this chapter, 'rock' is deployed as an umbrella term covering a range of popular music genres and styles, including 'classic rock', 'progressive rock', alternative rock', 'rock 'n' roll' and 'psychedelic rock'. The wide-ranging term 'rock' is useful for gathering data on the experiences, beliefs and perceptions of individuals with varying tastes due to the research placing much emphasis on the social uses and significance of popular music as opposed to more detailed analyses of what rock music actually is and how it is defined. The majority of my research respondents emphasized a degree of openness with respect to their musical preferences, and stated that they did not want to be labelled as fans of just one type of music—such as 'classic rock' or 'prog rock'. Thus, participants stressed their varying musical tastes and appreciation for a host of music genres and subgenres.

The longevity of rock music and the multitude of styles that have been categorized under the hyponym of rock mean that accurately defining what constitutes the genre is a difficult task. As Keightley (2001: 109) points out, 'Rock is a term that is instantly evocative and frustratingly vague.' Just as rock music is hard to define, the history of rock music is difficult to outline comprehensively. Scaruffi (2002: v) states, 'There is not one single history of rock music. There are several.' The histories of rock differ according to nations and individuals whose interpretations of rock music's roots diverge and depend upon distinct places and times, musical styles, genres and subgenres. Despite these variations, much of the literature acknowledges that rock music's origins lie within the emergence of rock 'n' roll, which developed as an offshoot of the blues, R&B and country music genres, during the late 1940s and early 1950s in the United States (Bradley 1992; Frith 1983; Keightley 2001; Peterson 1990). Artists such as Chuck Berry, Little Richard, Buddy Holly, Bill Haley and Elvis Presley brought rock 'n' roll music to the forefront of many people's lives in Britain and the United States. Inevitably, rock 'n' roll has since fractured into a huge variety of genres and subgenres. Moreover, in contemporary society rock artists who first began performing in the 1950s, 1960s and 1970s have grown older and provide a clear representation of the ageing process and bodily ageing. Indeed, rock artists and the fans that took part in this study have aged simultaneously; the ageing rock star presents a lucid reminder of the ageing process and the constraints of the ageing body for fans. This chapter explores the changing nature of ageing rock fans' behaviour during gigs and concerts, with a specific focus on embodied practices. The evidence demonstrates how rock fans have modified their behaviour, which has become increasingly refined and less concerned with the popular discourse surrounding the reckless 'rock 'n' roll' lifestyle, in conjunction with processes of growing older and bodily ageing.

Studying Ageing Rock Fans

My doctoral research (Gibson 2009) demonstrates that in contemporary society some members of former youth cultures have maintained their attachment to rock music and that older rock fans are able to negotiate an ageing identity with a music fan identity during their life course.[1] This chapter focuses on one facet of the broader research findings: the ways in which rock music fans experience the ageing body and how this shapes their relationship to, and involvement in, rock music culture. The theoretical framework of this chapter draws on a growing area of academic debate concerning ageing, the body and social theory. In recent years, the work of Featherstone and Hepworth (1991) and Katz (1996) 'has put the idea of the body at the forefront of analyzing social aspects of gerontology' (Powell and Longino 2001: 199). Until then, the social dimensions of ageing, which incorporate a focus on the body and embodied practices, were somewhat neglected in what can be characterized as 'medicalized' notions of ageing, mid-life and becoming old. This chapter contributes to existing understandings of embodied practices, and in particular how mid-life and 'older' people use cultural artefacts, appearance management (such as clothing and grooming) and space in response to ageing. The evidence demonstrates that, as rock fans age, they increasingly become conscious of bodily constraints and modify their behaviour at gigs; their reflexive understanding of themselves alters, which leads to fans engaging in practices such as limiting alcohol intake, remaining seated during gigs and adapting their physical appearance. Such practices potentially are more conducive to becoming old and bodily ageing.

This chapter draws on three data sources to illustrate how the ageing body has shaped a sample of self-defined 'older' rock music fans' involvement in rock music culture. The principal data sources were participant observation and both face-to-face and email interviews. Various documentary and Web 2.0 sources were also examined, including journalistic material, Web site articles, Internet message boards/forums and material used to publicize rock music events. The research participants do not form a cohesive social group, nor are they a typical sample of the population. Instead, their affiliation materializes from their enthusiasm for rock music, and their age categorization. Using a quota sample that takes into consideration age, gender and occupation, the research explored the opinions, beliefs and experiences of ageing participants in rock music scenes.

The method of participant observation was deployed in two ways: first, to observe how people aged over 30 behave, interact and use space in music venues, and second, to recruit research participants for in-depth qualitative interviews (both face to face and via email). Five rock gigs that catered to a range of rock music audiences were attended in the north-west of England. The bands and artists playing at the various gigs represented musical genres that can be placed under the all-encompassing label of rock. For instance, one gig featured a Pink Floyd tribute band, another featured a 'blues-rock' artist, a third featured a singer/songwriter rock guitarist, another presented a rock 'n' roll band that had been touring since the late 1950s and the fifth featured a 'heavy rock' band. All gigs included at least two performances, with the main artist/band having support from other rock music acts. The venues ranged from large-capacity, purpose-built sites such as Manchester Academy to smaller social clubs[2] that hosted a number of gigs on a weekly basis.

Participant observation at rock music events was combined with face-to-face and email interviews with 37 predominantly male fans from across the United Kingdom. The fans ranged in age from 32 to 58 years and were mostly middle class (a respondent's social class was assessed on account of his or her occupation).[3] The interviews covered a range of topics to supplement and confirm the observational data. These included the participant's life history as a rock music fan, the development of musical tastes, involvement in specific 'scenes' through the life course, behaviour and practices at gigs, intergenerational relations, the use of space, and gender/age differences at gigs. The data are used in this present chapter to illustrate how rock fans experience the ageing body and how this shapes their involvement in, and attachment to, rock music culture. The discussion begins with an overview of theoretical work concerning the 'mask of ageing', reflexive embodiment and postmodernity, integrated with empirical data to show how involvement in rock music culture allows fans to perform the ageing self, and to manage and negotiate potential conflict between supposed inner youth and the ageing body. Following this, the chapter explores the changing behaviour of older rock fans in terms of recovery, stamina and embodying age-appropriate discourses with reference to limiting intoxication and—a central tenet of this work—becoming more 'civilized'. Finally, an analysis of external displays of attachment to rock music culture, such as clothing and appearance, illustrates how older fans tend to internalize their commitment to rock music. The chapter concludes by arguing that long-term involvement in rock music culture allows fans to potentially blur traditional life-course conventions through their continuing attachment to, and participation in, rock music gigs. Nonetheless, the changing nature of embodied practices while at gigs, together with clothing and appearance, indicates the propensity for older fans to work within a number of established age classifications and thus revise their behaviour to correspond with the constraints of bodily ageing.

Ageing Bodies and Inner Youth: Reflexivity, Individualization and Postmodern Culture

In recent sociological analyses (e.g. Crossley 2006), there has been a growing interest in 'reflexive embodiment'. This term refers to 'the capacity and tendency to perceive, emote about, reflect and act upon one's own body' (Crossley 2006: 1). Thus, bodies are viewed as subjects and agents in the reflexive process of the construction of the self in late modernity. One's outward appearance is 'a means of symbolic display, a way of giving external form to narratives of self-identity' (Giddens 1991: 62). In consumer culture, individuals 'perform' the 'self': we are persistently on display and evaluated in terms of the success of this performance in the public domain. The body has become a central site for our sense of self in late modern societies. A further element of consumer culture is the focus on body maintenance, anti-ageing and a preoccupation with youth. In our everyday lives, 'we are overwhelmed with "gerontophobic" messages of youthful ideals: how to stay young, how to get older without signs of ageing, how to "stop the clock" and so on' (Oberg and Tornstam 1999: 629).

Recent work puts forward the notion that reflexive modernization has generated a socioeconomic and cultural milieu that allows for multiple, active ageing identities and differentiated 'cultures of ageing' (Gilleard and Higgs 2002) that blur and rework traditional age categorizations. During observation, it became evident that contemporary rock gigs—particularly those that involve performances from older and established rock artists/bands—provide a cultural space within which older fans are able to engage in performances of the ageing self. Through active engagement in rock music culture during adulthood, the older rock fan is able to engage in a modified version of traditionally youthful cultural practices. Thus older rock fans negotiate an ageing identity with their identity as a rock music fan, and—crucially—begin to rework traditional age classifications via their involvement in rock music gigs. The presence of older fans at rock gigs provides evidence of a blurring of life-course conventions: traditionally, gigs were youthful spaces and rock music culture was synonymous with youth. Nevertheless, the modification of embodied practices while at gigs—in accordance with growing older and bodily ageing—highlights the tendency for older fans to work within some established age categorizations, and consciously adapt their behaviour to fit with the constraints of their ageing bodies.

Postmodern theories of perpetual youth and a potential blurring of traditional life-course boundaries are consistent with a growing body of literature concerned with 'individualization' theories (Beck 1992, 1994; Beck and Beck-Gernsheim 1995; Giddens 1991, 1994) and a 'destandardization' (Brannen and Nilsen 2002) or 'de-differentiation' of the life course (Featherstone and Hepworth 1989). Recent analyses of the so-called baby boomer generation (Gilleard and Higgs 2002; Biggs et al. 2007; Phillipson et al. 2008; Leach 2011) examine the cohort of individuals born between 1945 and 1954 who were the first assemblage of people to grow up in consumer culture and have been characterized as having the potential to redefine traditional notions of retirement and old age. In consumer culture, identity is constructed and maintained via 'a series of "lifestyle-statements" that can be adopted or dropped at will' (Phillipson and Biggs 1998: 19). Consumerism permits a 'midlife lifestyle' for people that may be retained for as long as it is desired (Featherstone and Hepworth 1991; Turner 1995; Phillipson and Biggs 1998: 19). Thus, people in the mid-lifestyle (from about 30 to 60 years of age) can embrace a seemingly endless 'middle youth' (Featherstone and Hepworth 1991; Oberg and Tornstam 1999: 631).

Through the consumption of lifestyles that have at their centre a concern with bodily regimes (Giddens 1991), older generations can realize different 'possible selves' (Gergen 1991; Gilleard and Higgs 1996), and actively 'determine *how to be old*: what to wear, what to own, where to live, how to look and so on' (Oberg and Tornstam 1999: 631, emphasis in original). Thus it can be argued that older rock music fans engage in lifestyle practices that allow them to continue active participation in rock music culture and to situate themselves in midlife lifestyles as a potential continuation of their youth through the life course. Mark,[4] a 42-year-old historian, refers to maintaining his youth cultural practices during the life course: 'I go and see the same bands as when as I was young. I'll buy new albums but the music I listened to as a teen has stayed with me.' There is a degree of continuity and cultural commitment to this fan's involvement in rock music culture. Certain bands are seen recurrently in a live context as the bands and their long-term fans simultaneously age. However, as will become clear later in this chapter, behaviour and practices at gigs are modified as people grow older and become increasingly conscious of their ageing bodies.

Alongside optimistic accounts of ageing in postmodern societies, social theorists have argued that a serious tension exists among a person's subjective sense of self, his or her 'inner youth', and ageing bodily appearance. Such accounts imply that there is a 'mind/body dichotomy' (Blaikie 1999), with a marked variance between biological ageing and mental ageing. Featherstone and Hepworth (1991), for instance, propose the postmodern concept of a 'mask of ageing', whereby a gap exists between the external appearance of the face and body, and the inner subjective sense of personal identity. As one ages, this gap is thought to become increasingly significant. Thus, the mask of ageing thesis suggests that there is a disjunction between how old a person feels (state of mind) and how old they look (physical appearance). The mask thereby acts 'as a disguise which, layer after layer, increasingly conceals the "timeless" personality beneath' (Oberg and Tornstam 1999: 634)—or, as Blaikie puts it, 'the true—perpetually youthful—self [is] disguised behind old skin' (1999: 189). Turner (1995) purports that the paradoxical relationship between 'the subjective sense of an inner youthfulness and an exterior process of biological ageing' (1995: 258, cited in Oberg and Tornstam 1999: 634) is the central issue in the ageing process. 'With ageing,' Turner suggests, 'the outer body can be interpreted as a betrayal of the youthfulness of the inner body' (1995: 257).

A number of rock fans alluded to the theoretical concept of the mask of ageing when discussing how ageing had shaped their involvement in rock music scenes. Thus, on the surface, this provides empirical evidence of the tension between the ageing body and inner youth. Some fans maintained that their state of mind had not changed since their youth and teenage years. Indeed, several respondents recognized a degree of continuity regarding their sense of self. The following excerpt has been selected as exemplary in describing motivations for an ongoing commitment to rock music alongside recognition of the mind/body dualism:

> As I get older I realize that the only differences between the young and old are wrinkles, waistlines, hairlines and stamina, together with a refusal to accept poor service, quality and standards that once I wouldn't have batted an eyelid at (grumpy old man syndrome?). I still feel the same inside as I did when I was 20 so why should I stop? I didn't turn 40 and suddenly think—sod the Guinness, I want a cocoa (as a mature individual I now realize it's possible to drink both) (Rob, 43, chartered accountant, E).[5]

This fan professes that he feels the same psychologically as when he was younger. He argues that his state of mind has remained very similar despite obvious external displays of bodily ageing. Thus, while the body ages and shows clear signs of growing older, a person's mind may retain a sense of youthfulness. In a similar vein to Bennett and Kahn-Harris (2004), one could argue that youth is more analogous to an ideological category than a fixed phase of life. Nevertheless, it is important to make a distinction between how respondents *remember* their young selves (i.e. *claiming* that they feel the same now) and what they *actually felt* when they were young. Although

Rob maintains that he feels the same as when he was younger, the final part of his response (in brackets) indicates that he has a different attitude to life and people; evidently he has changed to a certain extent. The extract also highlights the notion of becoming more civilized, or refined, as rock fans grow older. Rob discusses his changing body with reference to wrinkles, waistlines, hairlines and stamina alongside repudiating 'poor service, quality and standards' in mid-life. Thus it is possible that he has grown increasingly concerned with how people behave towards him and what he puts inside his body (e.g. Guinness and cocoa) as he has aged; his social and physical expectations have therefore altered. In late modern culture, actively engaging in rock music gives fans the opportunity to perform the ageing self and negotiate the tensions between their inner youth and the ageing body.

Changing Behaviour at Gigs: Stamina, Recovery and Embodying Age-appropriate Discourses

In keeping with growing older and bodily ageing, fans have adapted their behaviour during gigs. Older rock fans tend not to engage in embodied practices such as energetic or frenzied dancing, or excessive alcohol consumption, which were once part of their rock experience. Such modifications illustrate how fans tend to work within certain established age-appropriate discourses and alter their conduct to correspond with the limitations of bodily ageing. Indeed, the changing practices of older rock fans indicate how they have become increasingly refined when participating in rock music culture. A number of interviewees described how their participation, with respect to what they actually do during a gig, has altered over time.

The majority of modified practices have occurred due to concerns surrounding ageing and the body. For instance, participant observation and fans' personal accounts demonstrate how most older rock fans tend not to enter into the 'mosh pit', and are more likely to sit down if there is available seating or stand near the rear of a venue or bar area, which differs from when they were younger. Mark described how his behaviour had become increasingly 'civilized' as he had grown older: 'I don't headbang[6] or dance, I stay seated best of all, as you get older you appreciate that.' This sentiment was echoed by the majority of fans who took part in the research. However, the only female respondent currently residing in the United Kingdom, Megan—a 31-year-old health-care worker—differed from the majority of male rock fans in stating:

> I like to be in the [mosh] pit and not stand back and watch from the back. I'm happy if I come home with bruises and a little sore, prof [*sic*] of a good gig . . . I do think now it is more of a stress relief being in a pit and venting aggression. I do find metal music kinda [*sic*] relaxes me.

For Megan, engaging in frenzied dancing and intentionally knocking into other fans in the mosh pit area are means of releasing antagonism and relieving tension. It must be noted that the fairly young age of this respondent—particularly in comparison to the majority of male research participants—may impact upon her penchant for entering the mosh pit.

Observation at various gigs highlighted the tendency for older rock fans to remain fairly quiet during live events and to be respectful of bands' performances by watching and listening intently to the music. If seating was available, the vast majority of audience members would sit down; thus, with respect to occupying space in a venue, the findings substantiate the work of Bennett (2006) and relate to the previous chapter in this book by Tsitsos concerning slamdancing and ageing. The older punks in Bennett's study legitimated their continuing attachment to the scene in two ways. First, with respect to occupying space in the venue, older punks tended to stand away from the immediate stage area, nearer to the rear of the venue and outside the mosh pit,[7] which can often be dominated by younger fans and is characterized by frenzied dancing. As Bennett goes on to note:

> The second situating strategy of the older punks was articulated through a form of discursive practice, whereby they positioned themselves as critical overseers of the local punk scene. (2006: 228)

Rock fans appear to have a mutual understanding that while it is acceptable to engage in loud conversations before and after the live performance, it is not conventional to talk or disrupt live performances. During participant observation at rock gigs, it became evident that audience members perceived they were at gigs to consume the music they liked in a live and intimate context. This was confirmed during conversations with older rock fans, since a number of them expressed disdain towards people who talked stridently, or behaved raucously, during a live performance. Generally, the gigs attended comprised audiences from a range of age groups, with the majority of older fans remaining seated or standing near to the rear of a venue, and behaving in a polite and respectful manner. If audience members were standing, they tended to gently nod their heads to the rhythm of the music but moved very little in general. Following each song, audience members would collectively applaud and often 'whoop' and cheer if they were particularly impressed. Although on occasion there were exchanges between bands and audience members, this interaction was consistently reverential. During one gig, a national artist with an acoustic blues/folk rock sound (signed to an independent record label) asked the audience whether his music was too loud. The crowd collectively replied, 'No!'; the performer then stated, 'Ah I see, so you're all old rockers then.' The evidence thus far demonstrates how ageing tends to impact upon a person's behaviour during gigs and involvement in rock music culture. Chris, a 43-year-old IT engineer (E) describes how his priorities and desires have altered in respect to attending gigs and his involvement in rock music culture more generally. In response to a question about whether his participation in rock scenes has changed, he replied:

> In a way yes, we all got older, had relationships, got married, had kids, came out the other side, older and sometimes wiser. Now we realize you don't have to hang around the stage door to get an autograph, your [*sic*] much more intent on finding a decent Italian restaurant and a decent pint.

This fan exhibits a somewhat 'been there, done that' attitude, and it appears that the live performance is an important but not an overriding or dominant factor in the experience of attending a gig. For this respondent, gigs are viewed as an element of the whole leisure experience of a 'night out', and locating somewhere pleasant to eat and drink are also important elements of a satisfactory gig and leisure experience.

The amount of alcohol consumed during a gig was mentioned by a small number of fans as a feature of gig-going that has changed. Descriptions tended to focus on how fans drink less alcohol than they used to during gigs or concerts. For instance, Mick, a 48-year-old library information assistant (F), recalled:

> I associate cider with gigs; cider got you drunker than lager. Now I drink bitter, not cider, I have drunk cider recently but it's that *American Beauty*[8] syndrome where you think you can handle it but it's a bit silly.

American Beauty is a 1999 film narrated by the 42-year-old male protagonist Lester Burnham (played by the actor Kevin Spacey). The film's plot centres around Burnham's 'mid-life crisis' and self-liberation as he becomes infatuated with his daughter's 16-year-old best friend, resigns from his advertising executive job, buys his dream car (a 1970 Pontiac Firebird), begins working out[9] and, significantly in respect to the respondent's quote, starts smoking marijuana. By referring to '*American Beauty* syndrome', the fan alludes to the notion of re-engaging in supposed youthful pursuits such as excessive drinking and assuming that one will respond to intoxication in a similar manner to when they were younger. Yet, as Mick notes, this is followed by acknowledging that such behaviour is perhaps rather foolish and non-conducive to mid-life. It is highly interesting that the film has become intertwined with the everyday discourse of the ageing fan, and used as a means of describing how people negotiate and rework embodied practices, such as alcohol consumption, as they grow older. David, another respondent (E), stated that having to get up early in the morning for work often prevented him from going to gigs. This, he explained, 'has become more important as I get older'. Evidently, the stamina-recovery process intensifies as fans age; this ultimately leads to a decline in the level of intoxication when participating in rock music culture.

Some research participants discussed matters concerning ageing, the body and feelings of ambivalence during their long-term involvement in rock music culture. Also mentioned was ageing as an inevitable and mutually experienced process between fans and rock artists. Fans discussed how they were prompted to consider, or were reminded about, the process of ageing as they witnessed their idols and favourite performers grow older. Older rock fans' experiences of ageing and the body are intertwined with representations of the ageing body from rock artists and other fans. An obvious and prominent example of an ageing rock star is Rolling Stones front man Sir Mick Jagger, whose career has now spanned five decades. Not only do artists such as Jagger provide depictions of the ageing body, they also represent how to grow old, what is acceptable in mid- and later life, and what is not. Jagger's body is considered to have aged well, as noted by a newspaper article in *The Times*, published on his sixty-fifth birthday, 26 July 2008:

> The grand master of heritage rock turns 65 today, and so officially becomes an Old Man. But he doesn't look it, even if his face does bear the odd wrinkle of a life thoroughly lived . . . he is the Duracell Bunny of ancient rockers. (Hamilton 2008)

Interestingly, the author equates wrinkles with having lived a full and consummate life, a rather positive evaluation of bodily ageing in which older rock fans may find reassurance. Jagger also exemplifies the notion of becoming more civilized as he has grown older, which is a central argument of this chapter. The article describes him in the following vein:

> At 65 he has become a spicy mixture of ageing bad boy, who once did unspeakable things with a Mars bar, and respectable English gent, with his holiday château and love of cricket. (Hamilton 2008)

Jagger's ageing persona is, according to this writer, perhaps akin to the older fans who took part in the present research: an amalgamation of his reckless 'bad boy' image with a more refined and gallant older self.

A number of respondents commented on feeling too old at gigs where the majority of the audience was younger—predominantly aged below 30. On the other hand, Clive, a 57-year-old IT consultant, explained how being a member of a primarily older audience had led him to reconsider attending gigs of particular bands:

> The first 'big band' (at the time) that I saw live was the Moody Blues. This . . . was at the time that they were 'at the top'. This was either 1968 or 1969. They still tour the UK every couple of years and I still go when they visit Manchester . . . However although I still go and see them I nearly stopped going a couple of tours ago because standing outside in the queue it was a bit like an 'old farts convention'—DEPRESSING!!—but such is life. They are still good to watch even though they are starting to fall by the wayside as a result of old age/illness. How are the mighty fallen (E; respondent's own emphasis)!

For Clive, being a member of an ageing audience and watching a band whose members are creeping into old age and facing deteriorating health has led him to feel disheartened and despondent about the ageing process. As fans and performers collectively grow older, some fans can encounter feelings of dejection. Older rock fans can experience a degree of sadness in witnessing their idols growing older and beginning to 'fall by the wayside', as Clive suggests. Clearly, the experiences of the sample of respondents indicates that there is a tendency for rock fans to feel various emotions concerning the ageing body during their long-term attachment to rock music culture. The evidence demonstrates that older fans often modify their behaviour at gigs and concerts as they grow older and face the inevitability of bodily ageing and changing priorities.

Cultural Artefacts and the Body: Changing Image and Appearance

Alongside changes in behaviour during gigs and concerts, older rock fans are also likely to adapt their style, clothing and appearance. Bennett's (2006) study of the continuing significance of punk rock for an older generation of fans outlines a number of issues of key significance to this phenomenon. His thesis stems from fifteen semistructured interviews and conversations with male punk fans between the ages of 35 and 53 years in the East Kent region of south-east England. Bennett also observed older punk fans at a number of gigs in the same area of the United Kingdom, using a similar methodological approach to the present research. Bennett found that older punks toned down their image despite this still having importance as a symbol of one's punk associations. For example, clothing and hairstyle were less visible signs of attachment to punk music and culture. This, Bennett contends, suggests that 'as they have aged their reflexive understanding of themselves as punks has become far more subtly articulated' (2006: 225). Furthermore, Bennett argues that this may also imply that the notion of 'allegiance' to punk changes with age. Thus he suggests:

> Rather than commitment being externally communicated through spectacular style, there appears to be a shared understanding among older punks of having 'paid one's dues' in this respect, the proof of commitment residing in the individual's ongoing, but matured, punk persona. In other words, from the point of view of older punks themselves, sustained commitment to punk over time has resulted in them literally absorbing the 'qualities' of true 'punkness', to the extent that these exude from the person rather than the clothing and other items adorning the surface of the body. (225)

Bennett's research highlights how older punks internalize the punk ethos as they age. It appears that identification with punk culture manifests itself in a more understated manner in comparison with when fans were younger. These findings are consistent with the experiences of older rock fans presented in this chapter, and theoretical work on bodily management in mid- and later life. Oberg and Tornstam (1999), for example, discuss how an embodied identity can be understood as 'a negotiated process whereby the individual actively and creatively draws upon cultural resources for making sense of who s/he is, who s/he was and who s/he might become' (630). Thus, older fans use clothing and other cultural artefacts related to rock music culture to make sense of their present, past and future self. Like Bennett's (2006) research participants, there is evidence of rock fans toning down their external image—such as haircuts and clothing—as they age. For example, most fans said that they had cut their long hair and some stated that they no longer wore items of black clothing and 'band T-shirts' to express their affiliation to rock music culture. This suggests that as rock fans age, their presentation of self and external displays of fandom are expressed in a far more understated manner.

During observation at gigs, dress and clothing tended to reflect a person's age rather than their attachment to rock music culture. For example, male rock fans tend to wear jeans with a casual T-shirt or shirt and trainers or shoes, as Pete, a 42-year-old journalist (F) commented:

> There's a point where you stop dressing tribally, it's when you're about 30. You progress in your 30s and become more assured of your tastes and dress how you want. You're more likely to wear expensive smart-casual clothing.

In further support of Bennett's study, it appears that this may indicate a shift in the way the notion of commitment to rock music alters with age. Moreover, there appears to be a reciprocal belief among older rock fans of having internalized one's commitment to rock culture, which resides in the fans' ongoing, yet matured, rock fan personae (Bennett 2006).

Nevertheless, in contrast to Bennett's (2006) findings, some older rock fans do objectify their commitment to rock music culture and particular bands through external image and clothing. Ben, a 39-year-old author, musician and blogger (E), described how he had long hair, mostly wore black and looked like a 'metal head'. He explained, 'While I don't wear metal T-shirts all the time it is clear that I am a heavy rocker.' Similarly, Richard, a 43-year-old chartered accountant (E) who plays cricket for a local village team and manages the local junior team, stated: 'I am usually berated for turning up to matches in Marillion/Hawkwind T-shirts'. Thus, while older fans may tone down their 'rocker' image, some continue to wear clothing associated with being a rocker together with other visible signs of their attachment to rock music culture, such as longer hairstyles, earrings and/or leather jackets. A small minority of older fans using certain cultural resources and bodily modifications to express their attachment to rock music culture was evident during observation at gigs. Such cultural artefacts and body alterations included clothing (e.g. leather jackets, band T-shirts, denim jackets and tight skinny jeans), piercings, hairstyles and tattoos. These were common amongst only a few, mostly male, older fans. Thus, perhaps the majority of older fans have internalized their cultural attachment to rock music, in accordance with Bennett's (2006) findings, and tend to express more mundane everyday articulations of fandom as opposed to dramatic displays.

Conclusion

This chapter has illustrated that rock music culture remains highly significant for many fans throughout the life course. Although older fans continue to participate in rock music scenes, they tend to modify their behaviour as they grow older. Key aspects of an ongoing commitment to rock music culture are: the changing nature of adult fans' behaviour during gigs; changing body image and appearance; and finally, feeling and behaving differently—despite claims by fans that they feel the same as when they were younger. This chapter presents support for 'individualization' theories (Beck 1992, 1994; Beck and Beck-Gernsheim 1995; Giddens 1991, 1994) and the 'mask of ageing' (Featherstone and Hepworth 1991) to a certain extent. However, it is clear that certain constraints of the ageing body do impact on participation in rock music culture, and that postmodern theories of perpetual youth fail to recognize the nuances and complexities in the lives of 'older' music fans.

The issue of recovery and stamina has been a key theme throughout this chapter. Bodily ageing and the ageing process have clearly influenced the types and nature of intoxication during involvement in rock music scenes. Whereas forms of intoxication including recreational drugs and alcohol vary from gig to gig, processes of recovery and recuperation tend to take longer during adulthood in comparison to the experiences of fans during their adolescent years. Equally, a number of fans referred to a reduction in stamina with respect to practices such as 'moshing', 'headbanging' or standing at gigs. As fans grow older, their health and physical fitness can affect and alter the practices undertaken during music events and gigs. This challenges the claim that some older fans believe they feel the same now as when they were younger, as evidently the ageing process does impact recovery and stamina at rock music gigs.

The presence of older fans at rock gigs provides evidence of a blurring of conventional life-course boundaries; traditionally, gigs were youthful spaces and rock music culture was synonymous with youth. However, this chapter has demonstrated that modifying embodied practices and altering body image and appearance during participation—in accordance with growing older and bodily ageing—signals a tendency for older fans to work within some established age categorizations and consciously change their behaviour to fit with the restrictions of their ageing bodies. While it is perhaps fitting to suggest that life-course boundaries are open to negotiation, symbolic and contestable, the experiences of a sample of older rock music fans highlight that bodily ageing impacts on participation and fans' behaviour reflects long-established age-appropriate practices and conventions.

Part III
Resources and Responsibilities

–7–

Dance Parties, Lifestyle and Strategies for Ageing

Andy Bennett

Emerging during the mid-1980s, the electronic dance music (EDM) scene presented as a new chapter in the history of post-war youth music and style. Characterized by new forms of DJ and computer-based music, a new synthetic drug, Ecstasy—or E[1]—and set within a culture of all-night dance parties—initially referred to by the British popular press as 'raves' (see Thornton 1995)—EDM quickly expanded into a broad church of local and global scenes (St John 2009). Alongside urban-based scenes, dance parties were also frequently held in greenfield sites. Typically deemed to be apolitical, in truth dance music parties often embraced highly radical political agendas. For example, the Free Party scene in Britain combined elements of the countercultural 'back-to-the-land' ethic with a Do It Yourself (DIY) aesthetic drawn from punk, and became a focus for widespread discontent with mainstream British politics among a disenfranchised sector of British youth (McKay 1998).

Sensationalized media representations of selected events and images associated with the Free Party scene completed the work of alienating the EDM scene through reducing it to a stereotype of trespassing, vandalism and drug orgies. Some twenty-five years on, the memories and cultural resonances of the dance parties of the early 1980s continue to have a significant impact on many of those individuals who were involved. Indeed, for many the attraction of the dance party has never faded, resulting in myriad strategies among ageing individuals to remain active in particular scenes.

Drawing on interviews conducted between 2003 and 2005 with dance party participants between the ages of 40 and 55, this chapter examines some of the lifestyle strategies created and employed as a means of ensuring a continued attachment to EDM and its primary spheres of consumption—the club-based or outdoor party event. Particularly significant is the way in which ageing dance party participants manage their involvement in dance music, as both organizers and consumers, within a range of other roles and responsibilities, including work and family life. As the chapter reveals, for many work, family and the dance music scene do not represent separate spheres of life. Rather, strategies have been developed that allow, where possible, for the integration of these apparently disparate elements into a coherent lifestyle (Chaney 1996).

EDM and the Spectacle of 'Youth'

During the 1980s and early 1990s, the popular media revelled in portraying EDM as an inherently deviant brand of 'youth' music. As with media representations of earlier incarnations of youth music and style—notably mod (Cohen 1987) and punk (Hebdige 1979; Laing 1985)—EDM was pilloried as a threat to social order due to its allegedly corrupting effect on youth. The problems inherent in such media treatments of EDM, together with earlier youth musics, have been analyzed and critiqued effectively elsewhere (e.g. see Thornton 1995). In truth, however, the extent to which the EDM scene was at any point in its history quintessentially focused around youth is open to debate. Indeed, as McKay (1996) notes, many of those who were involved in the Free Party scene of the 1980s had previous involvement in the hippie festivals of the late 1960s and 1970s, and were thus well beyond youth, in terms of biological age, when they became associated with EDM. Similarly, in her work on the more urban and commodified EDM scene of the mid-1990s, Thornton (1995) draws attention to the presence of older clubbers among the dance crowds she studied as part of her research. Thornton interprets this presence as a resistance to 'social ageing'. However, as will become evident in subsequent sections of this chapter, there is a clear case for arguing that what Thornton was witnessing was not a resistance to social ageing as such, but rather the beginnings of a broader shift in attitude towards the relationship between ageing, identity, leisure and lifestyle. In this context, music and associated forms of leisure, as these are consumed by older audiences, do not necessarily exist as part of an oppositional strategy to other, perhaps less enjoyed, aspects of everyday life. Rather, they are considered to be, and operationalized as, a broader lifestyle strategy and a means of re-presenting ageing identities in a time where (at least in a Western-developed social context) age is coming to be regarded as increasingly less important as a barometer for sensibilities of taste in music, fashion and associated forms of leisure and consumption.

Such a position also challenges another frequently posited observation regarding EDM. Thus, although much dance music scholarship rejects the tenets of the Centre for Contemporary Cultural Studies, subcultural theory in favour of more contemporary, postmodern-influenced writers such as Maffesoli (1996), at the same time many dance music researchers continue to focus on questions of resistance, linking these to ritualistic practices of hedonism and escape (e.g. see Melechi 1993). This aspect of EDM culture is said to be accentuated by the socioeconomic context of its origins—that is, post-industrialization and the concomitant emergence of a new leisure society. In such a context, casual labour and unemployment are argued to offer a relatively unstructured lifestyle, and thus more scope for regular participation in clubbing (Thornton 1995). However, to claim that all individuals involved in the EDM scene fit this category is something of an exaggeration. EDM has, in truth, attracted individuals from a variety of different walks of life, with a number of older EDM scene participants having professional career backgrounds (Bennett 2000).

Ageing and the 'Party' Lifestyle

Since its re-emergence though the cultural sociological work of theorists such as Chaney (1996), the concept of lifestyle—originally grounded in Weber's work on status groups—has been applied in a range of contexts to describe the ways in which individuals appropriate popular culture resources—typically objects, images and texts—in the reflexive construction of new forms of shared identity, or lifestyles (Bennett 2005). In more recent work on popular music and ageing, Bennett (2010, 2012) has developed the focus of lifestyle theory to extend across aspects of biography and memory as these are aesthetically operationalized as resources for the reproduction of individual and collective identities over time. As Bennett (2010) observes, a critical element within this process is the negotiation of the physical ageing process via reflexive engagement on the part of the individual with an alternative frame of reference for biographical development—that is, a process of 'cultural' ageing. The latter encompasses an understanding on the part of the individual that, although the ageing process and the biographical transitions that routinely accompany it are inevitable aspects of everyday life and the human condition, their bearing on the opportunity horizons and developmental pathways of the ageing subject are not absolute. Rather, they are managed through the reflexive lens of cultural ageing and, as such, tempered and balanced through an ongoing engagement with pre-embedded aspects of a lifestyle aesthetic as these orientate around aspects of taste, consumption and leisure.

In discussing their relationship to and participation in the dance party scene, older scene participants offered a number of interesting insights into the perspectives—both philosophical and practical—that age and responsibility had engendered. A common characteristic of those interviewed was that no one expressed regret at having grown older or felt that their ongoing involvement with the dance party scene amounted to a refusal to grow up. On the contrary, interviewees appeared to be highly invested in the lifestyles they had built for themselves, and typically perceived no inherent problems or contradictions in combining their lives as middle-aged adults, professionals, parents and so on with the dance party environment. Indeed, for many the ageing process had brought its own rewards, often articulated in terms of an ability to balance, and thus reap the benefits of, a range of activities and the concomitant expectations that these brought to bear on the individual.

Sustainable Fun

When discussing the relationship between work and leisure, and the importance of music within this, a number of my interviewees located the significance of dance parties in a context of what they referred to as 'sustainable fun'. This was particularly evident among interviewees from professional backgrounds. For such individuals, the relationship between work and clubbing embodies an obvious, and in many ways precarious, series of contradictions. Indeed, for those with professional careers and responsibilities, the practice of clubbing demands an altogether different form of commitment—together with a high level of self-discipline—in order that two very different lives can be managed and maintained. Speaking about his involvement in the London dance club scene—along with two friends of a similar age to himself—Simon, an international property consultant in his late forties, offered the following account:

> I was working in a nine to five job. So were my two friends. Y'know, we were pulling down good salaries and doing grown up things. But on Friday we were going out, and we'd go out from Friday night usually until Saturday evening, sometimes Sunday lunchtime. Just going all the way through. And then driving back, recovering on Sunday evening and then going back into work again on the Monday. And we would do that three, sometimes four weekends a month.

As noted above, Thornton has suggested that older adults are sometimes drawn to dance music and other 'youth cultural' scenes as a means through which to resist 'social ageing [and] resigning oneself to one's position in a highly stratified society' (1995: 102; see also Du Bois-Reymond 1998). At one level, the above account from Simon lends weight to Thornton's observation: his reference to 'doing grown up things' suggests that he and his friends simultaneously seek a diversion from such commitments through their involvement in the dance club scene. At the same time, however, further comments offered by Simon demonstrate a more complex interplay between his work and leisure life in which his relatively high-profile socioeconomic position was actually regarded as integral to the pursuance of a satisfying lifestyle. For Simon, his age and responsibility necessitated a rigidly adhered to series of checks and balances, particularly in relation to basic issues of health and well-being. While clubbing was a key leisure activity for Simon, it was also an activity that had to be carefully managed; clubbing is thus regarded as merely one component of a meticulously coordinated lifestyle project in which each element is as important as the next in the maintenance of 'sustainable fun'. This was clearly articulated in Simon's expressed attitudes towards drugs, and how these had changed over the years:

> You have to go one of two ways. You either lose it or you go into some kind of mark of craziness because, y'know, it does fiddle with your system. You have to step back and go, phew. There are two things I would never do, that's heroin and crack. But I've done everything else, sensibly. Because it enhances it. You know, it makes the journey more

pleasant. [But] it isn't a destination in itself . . . and those people who confuse the journey with the destination are the people who get lost along the way. And end up in rehab. And I don't intend to do that, because I've got a wonderful life. I can stick two fingers up at the establishment in the nicest possible way. I ask for nothing more.

According to Simon, another major factor in his successful management of clubbing activities and professional workplace responsibilities was the sense of continuity he experienced between his work and leisure spheres. A key element in this, suggested Simon, was the fact that his employer was also an experienced clubber, and thus both understood and shared Simon's commitment to the club scene. According to Simon, while dance music and clubbing were rarely a topic of conversation between them, the fact that both he and his employer pursued and managed their lifestyles in similar ways had created a strong bond of trust and collegiality between them. As Simon observed:

My boss can say he's taking time off and won't be back until Tuesday next week. I know exactly what he wants and how he's going to go about getting it. But he also knows that . . . because I party in the same way that he does that he can trust me implicitly in terms of presenting a professional face for the company on Monday morning when I'm running it while he's not there. And, for me that's fantastic . . . To actually have that as a thread running, not just through your social and your personal life but through your working life as well.

The importance of sustainable fun was also pointedly emphasized by older dance party scene participants involved in the organization of free parties. As with more commodified club-based dance party events, the typical representation of the Free Party scene suggests that this is sustained largely by those whose lives are not bound by the strictures of full-time employment, regular working hours, and the demands and responsibilities that come with this.

The following account, offered by Neil—a party organizer in his mid-forties—portrays an altogether different scenario. Like Simon, Neil suggested that, as an older dance party scene participant with work and domestic responsibilities to consider, the key to his success in and enjoyment of the Free Party scene depended on a continual balancing of priorities. Moreover, as Neil went on to explain, although arranging parties involved a great deal of commitment, the parties were in the final analysis only one aspect of his life and had to be managed in relation to other, more important commitments, including the coordination of several part-time jobs:

AB: How do you fit the rest of your life in around [organizing parties], because it sounds quite time intensive?

Neil: I think because it's teamwork . . . if I was doin' it all on my own, then I wouldn't be able to do it. I have two jobs that I do, an' I have them as my priority really.

AB: Your jobs are your priority?

Neil: Yes, my jobs are my priority [laughs]. The parties don't pay the bills, really.

Neil then went on to offer some candid observations on the difficult process of managing an ageing body in the context of a scene that places a high level of emphasis on physical exertion over long periods of time and during hours that normally would be spent sleeping. In talking about the extreme physical demands that his lifestyle placed upon him (particularly during the summer months, when he may be involved in the organization of several parties each month), and how these increased as he grew older, Neil described the strategy he had adopted for coping with this:

> You're stayin' awake all Saturday night and then getting' to bed, say midday Sunday. I might have four hours' sleep, wake up, 'ave something to eat and then go back to bed again. Try to get ready for Monday mornin'. An' you're just knackered really, so Monday can be a bit rough. So you 'ave an early night Monday to make up them two or three hours that you didn't get on Saturday night.

Accounts such as these are noticeably out of step with descriptions of the dance party scene that portray it as a space exclusively reserved for the exertion of youthful energy buoyed up by a 'twenty-four-hour' party lifestyle. Rather, it brings a new dimension to our understanding of the contemporary dance party scene and those involved in it. Assuming the 'rightness' and 'naturalness' of taking their lifestyle preferences forward with them into middle age, both of the ageing dance party participants discussed above show themselves to be confronted with a common challenge: how to adapt to the demands of the dance and party scene with an ageing body and the added responsibilities that age has brought with it.

Dance Parties as Family Outings

Another significant insight arising from interviews conducted with ageing dance party participants was the increasing tolerance towards the inclusion of families and children within the dance party scene. As original members of the 1980s Free Party scene have aged and the scene has diversified into a number of smaller—in many cases localized—subscenes, family involvement is becoming an increasingly normalized aspect of the dance party experience. This was evident in accounts offered by members of a small, close-knit dance collective based in the southeast of England that exhibited something of an open-door policy regarding the inclusion of children. In addition to being a necessary organizational element in the staging of dance parties for and by people who wished to involve their families, this informally observed policy of inclusivity was also considered highly desirable when contrasted with what was considered an overtly inscribed exclusivity among more youth-centred scenes. As one of the principal organizers involved in the collective explained:

> [A lot of] youth cultures focus on a particular genre to exclude other people. Hip hop is a classic example because it's quite narrowly defined [and] the hip hop guys listen to a

> specific type of hip hop and they all dress in a certain way, an' it's their scene. Whereas with ours, it's more open to anyone who wants to come along, regardless of what age you are, what clothes you're wearing, or the fact that you've got a couple of kids in tow.

Indeed, according to Brian—another member of the collective in his early forties—assuming they were properly supervised, attending dance parties could constitute a positive learning experience for children. Thus, he observed:

> You take your kids along [to dance parties]. An' I think it's a good education for them. Okay, sometimes you've got to explain stuff to them. An' sometimes people get a bit out of order, an' you've got to say 'excuse me, my kid's here, could you pack it in please?' But normally people are okay about that.

For Brian, there was no apparent contradiction between his role as a father and his interest and involvement in a local dance party scene. This was a view shared by many others involved in the collective. Having made the decision to incorporate the party scene into their lifestyle, rather than trying to find ways to keep this and their domestic concerns apart, collective members looked for ways to combine these two aspects of their lives. Moreover, it was clear from the accounts of other members of the collective that in including children in party events, considerable care was invested in ensuring that this would be a positive experience for them. For example, in discussing the outdoor dance parties organized each summer by the collective, Jake—a collective member in his early sixties—provided the following account of the child care and leisure activities built into these events:

> [Children] are never left. There's always one 'auntie'. But at parties people have got tents anyway, so they'll do their bit and then go and sleep. This is it, it always works. Kids, they have the best time. Especially if we fix up a swing for them. And we organize stuff, you know, adventure walks or trips to the beach . . . 'first one to come back with a limpet shell gets a prize'. That's what kids should do, y'know enjoy being children. Sometimes we organize midnight feasts, y'know, a bit naughty. But, nobody's getting them up in morning. So, they get up at 10.00 or so, bit of breakfast then off to the beach for the day.

This dance party collective's pride in its ability to organize parties that were inclusive of EDM followers of all ages, together with the emphasis on risk-free fun countenanced by a careful policing of drug and alcohol consumption and quick and efficient resolution of any form of antisocial behaviour, was core to the collective sense of identity. It was clear that age played a significant part in this. Age and accumulated life experience, it was suggested, were key ingredients in their particular style of party organization—and also acted on occasion as a convenient and effective deterrent against interventions from legal and other authoritative bodies. One particular incident concerning an intervention by the local police as the collective was preparing for an event was related to me on several occasions, and is described below by Neil, the aforementioned DJ and party organizer in his mid-forties:

> The police arrived and I think they were quite surprised to see a bunch of middle-aged people. So they just told us to make sure we were packed up and gone by 10.00 the following morning or something and that was it. An' the following week, the local paper had this story that read something like 'middle-age ravers told to leave by 10.00 am—an' they did!'

It is interesting how, in the above example, both the police and the local press viewed the fact of middle-age participation in a dance party as somehow novel and apparently out of place, a view that illustrates in turn how the common, stereotypical notion of the dance party as a youth-driven, drug-fuelled cultural space continues to dominate the popular imagination. As the following example illustrates, however, not only does the increasing involvement of middle-age participants in dance parties complicate such a perception, but it may also engender new forms of generational exchange within the dance party scene.

Party Parents as 'Role Models'

A recurring theme in interviews with ageing dance party participants was the way in which they felt an interest in dance music had shaped their relationships with their children. This, they argued, was not merely evident in terms of shared, or divergent, patterns of musical taste. Rather, it was claimed that music became a platform for the development of understanding between parents and their children. In his book *Two of Us*, American author Peter Smith chronicles how introducing his 7-year-old son to the music of The Beatles provided an avenue for them to talk about a variety of subjects and, in the process, ultimately brought the two of them closer together. Thus, concludes Smith, 'Just as the Beatles were a portal to other music, they were a portal to a friendship between me and my son' (2004: 192).

Similar accounts were forthcoming in several of the interviews I conducted with ageing dance party participants. The following account, offered by Lynne, a professional woman in her mid-forties, concerning her and her husband's acquired interest in dance music and its impact on their son's musical taste and music-making practice, is an interesting case in point. Lynne's observations are particularly interesting in that they reveal how, having initially been quite ambivalent about dance music and buying into media stereotypes in which dance was firmly equated with drugs and represented as a 'youth culture', she and her husband underwent a significant transformation of attitude and understanding when they were first exposed to the dance club scene:

> If anything I probably had some fairly negative views on [dance music] because this was when my husband and I had teenage kids. So we associated it with rave culture and, y'know, 'dangerous drugs' and things that we didn't really have any grip on. So

> [initially clubbing] was really more of just a social scene, y'know, with the local people going to clubs and things like that, so that was how we got to know the music originally. And then . . . I think it was in 1998, some friends of ours took us to a club in Brixton which, at the time, was kind of, I suppose, part of the underground dance scene that was going on in the South East of London. And [that] one night in a club completely altered our mindsets, it just changed us completely really, frankly, And eh, y'know you get a bit evangelical about it really, and all your friends have got to go.

As the above account reveals, having initially been quite apprehensive about dance music, primarily because of the negative effects they feared it may have on their children if they were exposed to it, through their own absorption in the clubbing scene Lynne and her husband subsequently revised their opinions about dance music. Not only did this refer to themselves and their relationship to the dance music scene, but it also extended to their children. As Lynne observed, one significant aspect of this was she and her husband proactively encouraging their son's interest in dance music and involvement in the scene as a DJ. Thus she explained:

> Our son, who's now twenty-one, I think we had quite a big influence in his interest in dance music, and his choice of dance music, and what he's now into. He's got into DJing and doing his own parties and things like that. And I think it would be true to say that that really came from us, from what we were interested in doing, the music and the clubs that we went to. And from our parties. Our son DJed-out at the parties that we did.

Later in the interview, Lynne was asked about her other children and whether they had also been inspired by their parents to take an interest in dance music and become involved in the dance party scene. Lynne's response to this question is of some significance in that it is suggestive of new dynamics at play around parent–child relationships and the purchase of musical and stylistic taste in the forming of such relationships. According to Lynne, in contrast to her son, whose embrace of the dance music scene resonated with Lynne's own acquired feelings towards dance music, she felt her daughter led a rather more conservative lifestyle, this being punctuated by more 'mainstream' musical tastes and night-time leisure preferences:

> My youngest daughter, who's nineteen now, she's always been into more kind of soul-based funky music, and not really . . . I mean she goes out to local clubs with her friends. But she's not into clubbing in any serious kind of way. For her, it's more about going out to a local night with a few friends, having a few drinks and you know, coming home about one o'clock in the morning.

Such a depiction is significant in that it suggests an interesting deconstruction of the once clearly delineated generation gap existing between parents and children, and the role of music in relation to this. Lynne's portrayal of her daughter would suggest that the accusations of conservatism once directed at parents by their children through the expressive platform of musical taste may just as easily be inverted in an age where parents are as likely as their children to opt for particular musical genres and associated lifestyle practices.

Conclusion

This chapter has explored the relationship between ageing, lifestyle and popular music as this manifests itself in the contemporary dance party scene. As the chapter has illustrated, not only is this scene now multigenerational, but those participants in the post-youth phases of their lives are proactively seeking strategies through which to balance their scene involvement with other activities and demands placed upon them by work, family and other commitments. Thus we have seen how an ethic of 'sustainable fun' is applied by some ageing dance party devotees as a means of imposing self-regulation on certain physical and consumerist excesses associated with the scene. Such self-regulation permits the ageing dance party participant to engage with the scene in a way that is satisfying to them while not losing sight of their everyday domestic and employment responsibilities. Elsewhere in the chapter, it has been noted how ageing brings about new perspectives on the dance party scene as something that should be inclusive in relation to family life and the participation of children. In working to foster such openness and inclusivity, ageing dance party organizers and those who attend their events impose new ethics of responsibility and conduct within the dance party setting, relating to issues of personal conduct and issues of well-being, health and safety. The final part of the chapter considered how such new intergenerational dynamics within the dance party scene work between parents and children, often fostering interesting examples of exchange and knowledge transfer in relation to scene involvement and modes of participation. In the particular examples used in the chapter, parents with an active role in a local dance party scene had actively encouraged involvement in the scene for their son and created an opportunity for him to acquire and develop his DJ skills. As the dance party scene acquires longevity, it is likely that scenarios such as those described in this chapter will become increasingly common. Like other examples considered in this book, the dance party scene has developed to the extent that its cultural presence and reach can no longer be examined and theorized as pertaining purely to its currency in relation to youth leisure and lifestyle.

–8–

Punk, Ageing and the Expectations of Adult Life

Joanna R. Davis

Introduction

In a subcultural vein, punk is understood initially in terms that are clearly marked and recognizable. It's the three chords and bratty lyrics; it's the Mohawk and the safety pins; it's anarchy and rebellion; and ultimately, it's rock 'n' roll incarnate: loud, angry and hedonistic. The distinctions between 'us' and 'them' are considered obvious. 'They' are clean-cut, demure, suburban, grown-up, professional and corporate. 'We' are young, angry, amateur, DIY and in your face. In this distinction, the binary opposition seems evident and appropriate. But the emergence of adulthood starts to disrupt these binary oppositions, and complicates punk notions of us and them. In this disruption, punk must be redefined in order to continue to make sense.

I examine here how various 'spokespeople' of punk engage in this redefinition. Whereas mainstream journalism tends to frame punk solely in musical and stylistic terms (and maybe with some consideration of the punk 'attitude'), subcultural journalism demonstrates how some 'old timers' reconceive of punk in terms that allow for more compatibility between punk scene participation and the realities of growing older. These terms draw attention to the constructedness of the binary opposition between us and them, and thus demonstrate the fluidity of punk as an idea. When conceived of in ideological terms, punk becomes more than one kind of music, or mere youthful rebellion; it becomes portable and adaptable—an approach to life rather than simply a phase of it.

'Subculture' as a theoretical tool emerged from studies of youth culture, both in the Chicago School (e.g. Whyte 1943; Becker 1963; Matza and Sykes 1961) and in the Centre for Contemporary Cultural Studies (CCCS) in Birmingham in the United Kingdom (e.g. Hall and Jefferson 1976; Willis 1978; Hebdige 1979; McRobbie 1991). In *Subculture: The Meaning of Style* (1979), Dick Hebdige offered an account that focused on the style of subcultures in theoretical and historical space. However, his work exemplifies what some scholars have criticized about much

CCCS work: that 'its treatment and interpretation of style . . . takes no account of the meanings and intentions of young people themselves' (Bennett 2000: 22).

Several scholars (e.g. Renshaw 2002; Kotarba 2002; Irwin 1977) have shown how becoming an adult is hardly a sign of the end of one's participation in music. Because of its focus on youth, the concept of subculture is not as helpful in making sense of the complex and subtle process of identity negotiation participants experience as they grow into adulthood. The concept of the *scene*, however, is more useful because it denotes the openness that adult cultural experiences would demand. Work by Bennett and Peterson (2004) explores the music scene as a space where 'performers, support facilities, and fans come together to collectively create music for their own enjoyment' (2004: 3). They use 'scene' instead of 'subculture' to avoid presuming that 'all of a participant's actions are governed by subcultural standards' (2004: 3) in reaction against a static mainstream culture. Drawing on Bennett and Peterson, I argue that the concept of the scene enables us to theorize how ageing members negotiate identity within the punk culture. As Bennett (2006) points out, a concept such as scene 'portrays individuals as more reflexive in their appropriation and use of particular musical and stylistic resources' (2006: 223). Thus, if punk is conceived of as a 'scene', it is fluid in a way that allows for examining this interplay between punk scene involvement and the realities of adulthood.

If we can shift to seeing 'adulthood' as a component of culture that people negotiate, rather than a static, narrowly defined stage of life through which people proceed in some predetermined order, we can consider how people's participation in or identification with 'alternative' scenes or lifestyles might play out in relation to their conceptualization of adulthood. Punk, then, is one 'alternative framework' within which people may make sense of challenges to their identity. Building on work that uses the concept of scene instead of subculture (Bennett 2006; Davis 2006), I consider punk here as a scene, and thus as more fluid and portable than punk as a subculture.

Data Collection

This chapter is part of a larger study of punk scene participants in a local scene, navigating the transition to adulthood (Davis 2006). In looking for a way to help make sense of these individual identity-negotiation processes in the context of the local scene, I struggled with how to depict the world of 'punk' and its contested meanings. If the punk scene provides an alternative framework for becoming an adult, what exactly is its ideology, and where could I find it? Specifically, was there anything out there that spoke to how older punks had managed or were managing the transition? I considered sampling punk zines[1] from around the United States, but because zines act against institutionalized journalism, this complicates the search for them, especially in any systematic way (see Collins 1999; Schilt 2003a). I was looking for places where punk ideology might be recorded, accessible to your average punk on the street; it might be something I could look at as evidence of what punk might mean, or how it is articulated, outside of the mainstream media. In the midst of an unwieldy search, I collected some back issues of the nationally (US) distributed punk zine *Punk Planet*, based on content descriptions that seemed to intersect with my interests: interviews with older punks, 'remembering' those who had recently passed away, and article titles that indicated 'growing up' or 'growing old'. I also included an article from *Maximum Rock n Roll* (MRR), the San Francisco-area punk zine established in 1982. Some years ago, a friend had loaned me a 1992 issue with an article on older punks who were still in the scene. The data come primarily from these two nationally distributed punk zines in the United States, *Maximum Rock n Roll* and *Punk Planet*, and their interviews with over sixty people who have at some point participated in punk music and scenes. They were primarily men, and most—forty-eight—of them were musicians, but many also worked in scene-related capacities, or in politics/activism, in ways that these publications deem compatible with punk rock. Though this is hardly an exhaustive or representative sample of the range of punk community publications out there, or even of these few particular publications, it does provide a window into the discourse over the span of a decade. What we have here, then, are 'ideological expositors'[2]—spokespeople for punk—even though many of them would or do reject such a label. The focus on these older punks as adults serves the purpose of demonstrating some central concerns of punk, and how it is defined and reworked to fit across time, rather than being clearly situated in one moment of a person's life. I coded these texts for quotes on getting older in punk, and then examined these in more depth to generate the specific themes of music, politics/ideology and lifestyle as frames for talking about what punk means in order to make it portable across life experiences. Here I focus on lifestyle.

These interviews provide examples of the discourse of 'how to stay punk', and the ideology of punk operates here as part of a lifestyle. Bennett (2006) found that some

> older punks appear to have reached a stage where punk is viewed as a 'lifestyle' (Chaney 1996), a set of beliefs and practices that have become so ingrained in the individual that they do not need to be dramatically reconfirmed through the more strikingly visual displays of commitment engaged in by younger punks. (226)

The older punks quoted here demonstrate this shift, focusing on how their history in the punk scene and their ongoing participation—such as it is—are incorporated into their ways of being. As such, it becomes a useful frame for interpreting and accounting for behaviour and choices through the transition into and throughout adulthood.

Lifestyle: The Inevitabilities of Adulthood

Incorporating ideologies of punk into one's life opens up a possible way to go about 'adulthood', and suggests the ways in which values and ideologies linked to traditional youth 'subcultural' participation can be translated or transported into one's life. Taking on punk as a lifestyle requires confronting the so-called inevitabilities of adulthood. I examine here three 'inevitabilities', and look at how each is accommodated, challenged and redefined: the abstract 'growing up'; career and money (financial realities); and partners and children (often conflated by respondents). Though it seems inevitable that older punks encounter these as culturally expected aspects of adulthood, there is no inevitable experience of them. Thus these 'inevitabilities' force older punks to remake their understanding of growing up in a way that incorporates their ideology of punk resistance that they have so carefully cultivated in earlier years. I conceptualize accommodation here as a certain level of acceptance, where some of these aspects are taken as inevitable—and even regarded as positive on a certain level—ways to grow or accept the changes that happen with getting older. The line between challenging and redefining is a fine one, but I conceive of challenging as the resistance against a given definition, while the process of redefining is an attempt to give it new meaning, in addition to merely pushing against what is presented to you. As each of these 'inevitabilities' is a potential site of resistance, the ways in which each is redefined show how punk ideologies can inform lifestyle choices.

'Growing Up' in the Abstract Sense

The abstract idea of 'growing up' is a powerful frame for understanding the changes a person might experience in their twenties and beyond. One aspect of confronting 'growing up' is accommodation. For all their resistance, older punks ultimately understand that there is a level of accommodation that has to happen; in some ways, that accommodation helps to continue to resist the things that *are* resistible, while accepting certain changes. Joey Keighley, aged 35 and formerly a guitarist for the (now defunct) band DOA, speaks with the experience of having been involved in punk for more than a decade, throughout his twenties and into his thirties. He continues to play music with a new band, and explains his connection to music and the scene in this way:

> The whole thing was ongoing in the first place, and bands, just like friendships and relationships, never have a prescribed period of time, just as your life doesn't . . . I grew up, although your definition of that changes all the time. What made sense at 25 is different from what makes sense at 35. (*Maximum Rock n Roll* 1992)

By conceiving of his participation—including the friendships and relationships in his life—as 'ongoing', with a power outside of the individual, and naturalizing the process ('I grew up'), he demonstrates the usefulness of having a degree of accommodation, of not fighting himself—or forces that seem natural and external to him—too much. But in his case, this is not necessarily synonymous with growing *out* of the scene; although he has become a father, and plays slightly different music in his newer band, he envisions his current participation as a reasonable next step as an adult, but also as a punk. Thus his conceptualization of punk as ongoing makes room for growing up.

Though older punks may accommodate the *fact* of aging and 'growing up' to some extent, they can still find ways to *challenge* a normative concept of adulthood. The mere act of 'not growing out' permits living this challenge to that 'other' definition of adulthood. Dick Lucas, singer for the band Citizen Fish, sees this as a constant struggle:

> Everybody is struggling to make sure they stay where they are, or they've given up struggling and gone into the gray realm of otherness and normality. More normality begins to creep into the social life of being 30, and you have to keep tabs on where you are, and only by doing that can you keep tabs on where you actually will be (*Maximum Rock n Roll* 1992).

Though he clearly sees others giving in to normality, for Lucas, when those moments of normality 'creep in', each one is an opportunity to resist or challenge the blind acceptance or accommodation of it. Lucas sees this increasing contact with the normal world as a natural part of getting older, and though some people cross over to that 'gray realm', he does not translate that into a requirement to let it become his own 'normal'. By keeping 'tabs' on those moments, he can challenge some aspects of growing up.

The simple fact of other people still staying involved in the scene can be reassuring to people—they show how both punk and adult can be redefined to be compatible. In the introduction to his interview with musician Milo Aukerman, *Punk Planet* writer Joe Meno says of Aukerman, 'Speaking with him about growing up punk and facing the equally strange world of adulthood helped to reassure me that getting older does not mean giving up' (Meno 2004: 74). The conflation of 'growing up' and 'giving up' can be such a core fear for many younger punks that any proof that giving up is *not* inevitable is helpful. For Meno, Aukerman is one whose path appears to have embraced both his punk music and his adult life, demonstrating that these are not oxymoronic ideals.

Part of the redefinition of punk as an adult involves redefining that 'other' adulthood. Mike Watt explains how he sees the 'normal' path of male mid-life:

> I'm a 46-year-old punk rocker now, so this is *intense*. Where does a middle-aged man really sit in society? If you look at Squarejohns [referring to an average 40-something man], he's supposed to be set up in his middle-class thing with a family, getting solidified in some kind of vocation or career—this is the way it's supposed to be. The reality is that a lot of cats at that age are having *huge* crises. They feel empty. They might have these material things, but they're bored stiff with the routine. All these things they followed by rote in order to be 'happy' and 'an important part of society,' they feel lacking (Ziegler 2004: 46, emphasis in original).

To Watt, it is clear that the approved path does not actually hold much appeal, outside of the fact of its approved status. He sees that option as one that is deeply unsatisfying, even to those who have chosen it—further justifying his own choice not to take it. Instead, as he defines his chosen path as a '46-year-old punk rocker', it is not empty or lacking, and it does gratify him and operate for him in a way that continues to be compatible with his youthful ideals. But this also does not mean rejecting the fact of his current age, and all the things that come with it. He explains that it is a perspective, not a rejection of being a grown up:

> [W]hen you're younger and you have to find out who you are, you do have to get a little over-bold to get some nerve up. But the things is to let go a little bit, too . . . Like what Perry [Farrell] told me: 'Keep the child's eye.' He didn't mean be naïve, he meant you should try to keep the ability to still find wonder so things can still trip you out a little bit, make you step back, and size up the whole deal again. It's not about pretending, being infantile, or some shit—you're still an adult woman or man. (Ziegler 2004: 47)

For Watt, maintaining a sense of wonder or curiosity—both of which he sees as youthful traits—is compatible with being grown up. In his definition, it is 'authentic' in the sense of not pretending, or affecting some inappropriate identity or persona, but embracing the 'realness' of one's actual age, and finding a way to make that fit with the perspective gained from a life in punk rock.

In fact, those who do not 'grow up' are subject to criticism from peers, which can constitute another form of inauthenticity. Jello Biafra notices his changing attitude towards some of his fellow punk scene members:

> Another downside of what you call being an elder statesman is getting very fed up with the immature side of . . . seeing people turn up with the same old hang-ups I've seen come and go for 10 or 15 years, I find myself less and less patient with. Everything from people who are dishonest with their friends and relationships, or people who are really hardline 'more political than thou' for a year or two and then turn around and become a yuppie or a wimpy pop singer or something. (*Maximum Rock n Roll* 1992)

Not growing out of one's 'hang-ups' or immature behaviours, and the hypocrisy Biafra sees as inherent in the inconsistency of drastically changing political beliefs, are signs of an inauthenticity that he cannot accommodate in his life. In response to the interviewer calling him an 'elder statesman', he asserts his authority to redefine the meaning of punk as an adult to include emotional maturity as well as some level of consistency in his ideologies.

For many older punks, there appears to be a crucial stage at which they confront challenges to their ongoing participation, and make the decisions that will determine the path they take from that point forward. That is, they reach a point where they accommodate, but do not challenge or redefine. Tommy Strange explains:

> I reached that crossroad at around 24–25, a time when my peers were graduating from college, getting married, and some people changed a real lot. At 30, the people that I'm hanging around with . . . have gone through that and are pretty much lifelong misfits at this point. (*Maximum Rock n Roll* 1992)

Encountering those different markers of adulthood does not inevitably mean that one does 'change a real lot'—but it might. For Strange, though, getting through that stage and maintaining one's 'misfit' status sets a person up for a lifetime commitment to the role. For him, the meaning of adulthood includes this redefinition that includes being a misfit. Similarly, Mark Andersen, a 33-year-old activist, explains:

> I knew it was a lifetime thing, the only way it's going to make any difference. You have to set yourself up to go for the long run . . . if we understand that it's a long term thing, then we're able to roll with the punches a little more and get up off the floor. (*Maximum Rock n Roll* 1992)

For Andersen, this approach makes a person more resilient, more able to stay punk regardless of the challenges to it that might arise. Though Andersen suggests that he took this lifetime approach at an early stage in his scene involvement, he and Strange both suggest taking a long-term view—perhaps at that 'crossroad' where ongoing participation is challenged by various rituals of adulthood. If one looks at it as a phase, or a finite stage of life, then it will have more rigid boundaries—and is thus more likely to break, to not hold under whatever pressure comes up. But if one sees punk as a lifelong way of being, or a life's work, then one is better able to absorb the challenges that come up and continue on with renewed commitment to the punk ideology that shapes one's life.

Career and Money

With the financial realities that go along with adulthood—such as supporting children, paying for housing or saving for future security—most older punks maintain that one must have some sort of career to meet these needs. Some of them may idealize a music career as the best way to do this, but not necessarily. Many punk musicians note how hard it is to maintain their artistic integrity and to make a living with their music, and this is at the heart of every 'sell out' story. Janet Weiss of Sleater-Kinney demonstrates this sense of accommodation to the music career as different from other careers: 'I think I've always known that no matter how successful your band is, there's going to come a time when you have to get a job again' (Ryan 2004: 40). The music is not the 'job' that other kinds of work will be, and she still has a sense of inevitability about that reality, despite the relative success of her band. In this sense, Weiss accepts or accommodates the primacy of financial concerns in her career choice. But part of the problem can be the assumption that a 'straight job' is inherently incompatible with maintaining an active music fandom or creativity. Jim Testa, 37, publishes a zine and works by day in insurance. Like others who have such mainstream jobs, he experiences that tension between the different worlds he inhabits, but insists that maintaining a position in both is not impossible. When asked whether he feels 'a dichotomy between your straight life and punk life', Testa says, 'I just shift gears when I need to. I'm pretty comfortable with it' (*Maximum Rock n Roll* 1992). Having been in this position for some time, Testa has become accustomed to it, accommodating the financial reality of needing to have a 'real-world' career, while maintaining his sense of himself as punk. Metal Mike, 40-year-old guitarist and singer for the Angry Samoans as well as an accountant, experiences a bit more instability with this need to shift gears, explaining, 'My lives are separated quite clearly, with the result being mild schizophrenia. I can see myself going into work and saying, "Hey, look at my new record, *Slave to My Dick*"' (*Maximum Rock n Roll* 1992). For him, the tension is a bit more acute, as his stereotypical musician self and accountant self usually are expected to inhabit different worlds. But he still chooses that schizophrenic experience in order to maintain his music life *and* accommodate the inevitability of needing to have a nonmusic career.

The resistance against the typical career is present both for those who have straight jobs and those who do not. One might say that Metal Mike, in spending his nonwork hours making an album with a title like *Slave to My Dick*, is expressing his resistance to that reality in an indirect way. He does not indicate whether he, like Jon Von Zelowitz does below, introduces some level of resistance into his self-portrayal at work:

MRR: You're a computer programmer in your 'real life', and does that ever cause schizophrenia?

JV: Sometimes I feel kinda weird telling people that, for fear that if they don't know me very well they might think I'm a weirdo. And some people at work think I'm pretty weird, but I don't care. (*Maximum Rock n Roll* 1992)

In both aspects of his life, Von Zelowitz experiences some dissonance, but given the possibility that his co-workers might find him 'weird', he expresses some resistance through his claim not to care; he worries about how fellow punks might interpret it, but not fellow computer programmers.

In attempts to redefine the meaning of careers and financial necessity, older punks seek to frame how they choose and enact their careers as compatible with ongoing punk identities. For those who stay involved in music, they must couch their choices in a frame of 'responsibility'. Watt describes the musician's life as one of self-motivation:

'Not trying to say I stayed a kid all my life—in some ways, I had to be a lot more responsible. I don't have a boss telling me what to do' (Ziegler 2004: 46). Although not being told what to do is often a rebel cry against the pulls of mainstream careers, here Watt frames this freedom as a potential difficulty of the musical career, thus making his choice appear more responsible, in that it requires self-discipline in the absence of a traditional boss. His career, then, is a harder choice than taking the mainstream, job-with-boss path that most people tend to choose.

This kind of choice calls for a balance between the integrity that punk demands you maintain and the financial realities of self-employment. Bruce Pavitt's explanation of the tensions and choices he faces running the independent label SubPop demonstrates this:

> You believe that you and your friends and community can change things, the way people look at things, the way people relate to each other. You have to have something like that if you stick with it, but at the same time you have to pay your bills and be more business-like and responsible about things. It has to balance out. (*Maximum Rock n Roll* 1992)

For Pavitt, it is necessary to cling to the optimism, the belief in change and the usefulness of fighting the system, in order to keep the motivation to run the business in the face of all the financial difficulties that might entail. By reconfiguring their career paths as legitimate, adult choices, and simultaneously displaying continued adherence to punk ideals, Watt and Pavitt demonstrate the redefinition of career that helps them to stay involved in punk as a lifestyle. By defining their career choices in terms that are used in more mainstream work, Watt and Pavitt present themselves as adult in that sense—that is, they are redefining playing in a band or running a label as adult jobs. These activities may have begun as leisure but have become their work, and this redefinition had to happen for them as they reached adulthood to legitimize their continued involvement in the face of accountability to normative expectations of adults.

Children and Partners

Becoming a parent is a strong pull for older punks to accommodate some aspects of mainstream adulthood. This still doesn't mean that they necessarily lose touch with their scene-related ideals, 'sell out' or become absorbed into the system. But when it comes to making room for a child, for most folks the alternative lifestyle *has* to get flexible. As Ian MacKaye points out, 'life and children, those are *not* puzzles and they are *not* conundrums—they are *reality*' (Sinker 2004: 12, emphasis in original). By framing parenthood as 'reality' rather than 'puzzles', or something that must be figured out, MacKaye validates any sacrifices or changes a person might have to make to put their parenting in their lives as they see fit. The financial, moral and legal responsibilities of parenthood make this a weighty issue.

At the same time, older punks can explain their accommodation of parenthood in ways that completely normalize it, just as any 'mainstream' new parent might. They tend to couch it in positive terms, focusing on what it has added to their lives at least as much as on what challenges it has presented. Keighley explains this about having a family:

> It's helped me be more aware of what's going on in the world, having an even greater stake in trying to preserve the good parts, and it's a financial reality, too, when you have kids that are depending on you. I've changed, but haven't wildly changed. (*Maximum Rock n Roll* 1992)

Keighley doesn't deny undergoing some changes as a result of parenthood, but he frames them as primarily positive ones.

An even more explicit sense of the positive choice is evident in Ruth Schwartz's explanation of how her experiences (primarily running a distribution company called Mordam Records) in the scene influenced her decision to have a child:

> I was already feeling alienated in the punk community. I was sick of spending hours and hours in clubs and feeling that I had wasted my time, that I hadn't experienced anything worth all that time. The bands weren't that good, the experience wasn't that great, and I would've been happier doing something else. It was out of that feeling that I decided that if I was going to have a child, this would be the time to do it. I don't have any other oats to sow, and couldn't feel like I was being left out of something because I was doing something else. I was ready to do something else. (*Maximum Rock n Roll* 1992)

For Schwartz, the lack of enthusiasm she was having for her life as it was in the scene was evidence of her readiness to do something else. At 31, she continues to run a scene-related business (nine years later, in 2001, *Punk Planet* reported the perseverance of Mordam under Schwartz), but her parenting has proven her earlier conviction that having a child was 'the most important thing you could do'. She didn't find her work in the scene to be as important as she imagined raising a child would be, and was happy to move on to that stage of her life. In this way, parenting was the natural progression for her in her late twenties, with a partner by her side. Thus, to accommodate parenthood into her scene-related life made sense, and appeared to be an obvious and desirable step. Her work grew out of her youthful scene participation and became a job (redefined, like Watts and Pavitt, quoted above), and thus she balances it like any working parent might.

By contrast, some older punks resist the assumptions that adulthood *must* bring parenthood. Leesa Poole highlights the sense of choice around parenthood by making explicit what she thinks the requirements of parenthood should be—things that she is quite honest about not being willing to do at this stage in her life, or perhaps ever:

> I don't think you should have a kid unless you're ready to put the kid first and give it what it needs, and you need a lot of money and a lot of energy and a lot of attention to do that. I don't want to have to go find a steady job to support a kid or worry about who's going to watch my kid while we're touring Europe for a month. (*Maximum Rock n Roll* 1992)

For Poole, the punk scene—and the nature of touring—are not compatible with her ideas of what parenthood should entail. Although the scene's incompatibility with this notion of 'good' parenting is rooted in rock 'n' roll's masculine and historically misogynist tradition, she is resisting the expectations that she should become a mother by asserting a definition of parenthood that is incompatible with the scene she knows. In doing so, she resists both the overarching expectations of adulthood as parenthood, and those specific to women, along with the differential social power and vulnerability that are considered 'natural' to motherhood.

So how can parenthood, or traditional family life, be redefined to work differently with a person's punk preferences? Some older punks reconstruct parenting as an opportunity to resist and 'see' the world in new ways, which enhances their commitment to their punk ideology or lifestyle. Having built part of his musical reputation on explicit resistance to adulthood, through songs like 'I Don't Want to Grow Up' and a tenet of 'Allogistics' asserting, 'Thou shall not commit adulthood', Aukerman has found that parenthood presents this opportunity. He explains, with regard to that tenet:

> That's one that, try as you might, you can't really fight. That one really relates to your overall view of the world. I still try to maintain a youthful outlook on life, but it gets harder when you have kids—or maybe it gets *easier* when you have kids because you can really feel youthful through your kids. There are many things about parenthood where you can feel the gray hairs emerging on your head, but at the same time there are aspects that make you feel more alive, more complete. (Meno 2004: 78, emphasis in original)

Many people get involved in punk, or any music scene, as a way to feel more alive or complete, and for Aukerman parenthood now provides him with the chance to do this, despite the potential trials and constraints that come with it. His children are still young, so it remains to be seen how it might play out, but Dale Stewart has found that now his daughters are young adults, music provides another way to connect—a way to define the parent–child relationship:

> I look in the mirror and don't necessarily see a 41 year old man. I was still being carded at 7–11 up to 10 years ago. I've always looked younger and felt younger. My 2 daughters, who are here tonight, are 16 and 18. About an hour ago we played 'Too Drunk to

> Fuck' and 'California Uber Alles' and all of us were sitting around grooving on it. We had some other things in there too, like Roy Orbison and Iggy Pop. (*Maximum Rock n Roll* 1992)

For Stewart, punk can be a family activity, or part of a family culture. He and his partner play in a band together, and his daughters have grown to appreciate the music that is meaningful to him and to them in such a way that parenthood can be redefined as explicitly compatible with punk rock. Furthermore, their participation demonstrates that parenthood can become a vehicle for perpetuating punk ideology, and by passing it on to the next generation, or sharing it, the age-specific aspect of punk resistance can seem irrelevant.

In confronting 'domesticity' and parenthood, many older punks may naturalize certain aspects of it while still challenging whether or not it is inevitable, and redefining *how* to parent in ways that may be more compatible with punk rock. By framing parenthood as something that enhances their commitment to punk rock, or extends their youthful resistance, these older punks find a way to make parenting compatible with an ongoing punk identity. Each of these aspects of adult life—growing up, careers and family life—*can* be a site where older punks become *former* punks. But as the above quotes demonstrate, they also provide opportunities to *redefine* the meaning of adulthood in order to maintain a punk ideology and identity.

The Self as Consistent Across Situations

Through confronting these 'inevitabilities', many older punks develop a sense of self that is consistent across situations, linked to the self as punk, as musician, as creative—any identity that allows them to see their connection to their youth scene involvement as still present in their current lives. This self, then, is seen as stable across different contexts—whether they are 'within' or 'outside' of scene life—and punk is more adaptable to it. If punk and self are indivisible, it becomes quite clear that punk will be *how* one lives one's life.

Punk is frequently defined in terms of being or seeking truth, and a true self. Punk authenticity (as we might also see in other music forms, and other subcultural lifestyles) often comes down to being 'oneself'. One of the members of the original 1970s New York City punk scene, Joey Ramone, demonstrates this credo:

> I've grown a lot on the inside and continue to grow on the inside, and see things a lot more clearer than I did before. The game is just to grow . . . You gotta stay in touch and intact . . . Just stay true to yourself, that's my advice. (*Maximum Rock n Roll* 1992)

This is common advice from all these folks; therefore, a credo of punk is to stay true to oneself. This makes room for doing all sorts of different things, as long as it can be claimed to be true for oneself.

As the person who started *Punk Planet*, Daniel Sinker experiences the need for his work to stay relevant to who he is in the moment. In his late twenties, he explains to an interviewee that:

> *Punk Planet* started when I was 19 and what was interesting to me at 19 isn't as interesting to me now. If I'm going to continue to work on this, it has to be interesting to me. (Sinker 2003: 60)

Though his life's work, as it stands, is clearly scene- and punk-related (as demonstrated in the title of his zine), it is more important that it continues to matter to him; presumably, if he couldn't continue to make it something that slotted into his sense of self, he would cease to do it. But as with any vocation or career, we may become more defined by it with age. Carrie Brownstein of Sleater-Kinney comments on this trajectory:

> When you start out, you see yourself as having multiple identities, and most of your friends see you as having multiple identities. I would imagine it'd be the same with any career—you're a few years into it and suddenly that's who you are, whether you're in a band or in retail. (Ryan 2004: 40)

She likens the musician's career to any other, where the issue of identity and the career can be seen as universal. Her comment, along with Sinker's, naturalizes this process, and accommodates it to some extent.

But this can be constraining as well. All of the so-called legends are purported to push against it—MacKaye, for example, 'makes repeated references to feeling "hemmed in" by aspects of his legacy' (Sinker 2004: 30). Although we might understand why he is interpreted or understood first through his music and label-related activities, this resistance comes in part out of the suggestion in the idea of the 'legacy' that he should not change, that he should always be Ian circa 1982, or Ian circa 1989.

Aukerman has taken another route: he has actively pursued—and prioritized—his job as a scientist. The band named an album after his choice, *Milo Goes to College*. He explains this in terms of self-fulfilment that punks can presumably understand, saying, 'I just have other aspirations . . . I just can't be a professional musician' (Meno 2004: 76). He elaborates:

> I never really considered myself a punk scenester. I'd go see bands, but I never had an affinity for the punk scene itself—I just wanted to check out the music. What I did have was an affinity for academia and in the end, I feel like I made the right decision. At some point the draw of academia was just stronger than the draw of punk rock. (Meno 2004: 78)

Aukerman found his 'true self' in a world outside of punk, and claims legitimacy in this feeling. Despite this, he is still recognized by many fans, and in this case the *Punk Planet* writer, as an important older punk. He is not a pretender to the 'real punk' or 'real scenester' claim that some would feel compelled to make, but perhaps gains more legitimacy in the scene, in some people's eyes, than if he were to make false claims to such labels. In this case, the claim to a consistent identity becomes the thing that justifies or validates labelling him as punk.

Conclusion

Even when punk is defined as a kind of music, or a particular sound, older punks identify the meaning of punk more (or at least as much) in terms of how it functions in their lives: to connect with other people, to express emotion, to explore their creativity. Defining punk as a political or ideological stance helps to construct it as something that can be incorporated into one's approach to life, part of the beliefs and practices of the punk lifestyle. If the self is an ongoing construction—and in punk, authenticity emerges through accountability to self—what is the role of ideology and philosophy in the transition to adulthood? Coming to embrace change and paradox seems to be central to the experience of becoming an adult, and thus resistance must be reframed as a way of being, not necessarily a clear-cut rebellion against some nameable force.

We still have to function and get by in the real world. Fervent optimism gets sober, and the meaning of punk is reconfigured to make sense with the realities that come with adulthood. Although parenthood, for example, is not truly inevitable, it is one common responsibility of adulthood. The consistent self is constructed in an ongoing manner to accommodate—because on some levels one must at least acknowledge (which may be a form of accommodation) the ways in which one is constrained by structure; it is a way of acting against, of maintaining a resistant definition in the face of, structural forces. Some of these structural forces are welcome; some operate as privileges for some people. However, the necessity of contending with them calls for active negotiation of the self to make sense over time in the face of these constraints.

Articulated here as an exercise in complication by dozens of older punks, punk becomes something that can be incorporated into one's adult life. By complicating the definition of the punk lifestyle—which is then more broadly defined to accommodate adulthood—punk becomes compatible with the lifelong endeavour; it becomes a philosophical approach to living. In its complication it becomes portable; it can *fit* something besides youth, and narrow definitions of clean, obvious distinctions between youth and adulthood. As such, it helps us—sociologically—to understand how that transition to adulthood *is* a culturally constructed social phenomenon; does not have clear, obvious steps and trajectories; and need *not* be taken for granted as having some simpler, obvious meaning. Perhaps there are trends and tendencies, but we must remember not to assume that this makes them 'real', 'true' or anything more natural than the alternative.

–9–

Alternative Women Adjusting to Ageing, or How to Stay Freaky at 50

Samantha Holland

Can I become a different being while I still remain myself?

Pearsall 1997: 3

Introduction

In 1997 and 1998, I interviewed twenty women who identified as 'not traditionally feminine' for my PhD research; the work was later revised and published as a monograph (see Holland 2004). My initial interest lay in the absence of adults in subcultural literature, whereas from my own experience I knew that people in their twenties and thirties didn't suddenly grow out of the alternative identities they had maintained since they were teenagers. Although I had expected to hear stories about how the women resisted femininity, instead they told me about their concerns about ageing and the issues of being both a woman and a 'freak', and the issues they faced in being both. I found that the participants employed various strategies to balance being 'freaky' enough with being feminine enough—for example, 'flashing' their femininity (2004: 44–50), toning down (2004: 127–32) and placing themselves in opposition to the category 'girly' rather than against femininity per se (2004: 148–9).

Since then, I have been asked repeatedly whether I had considered carrying out further updated interviews with the same women. It appealed to me, but it never seemed the right time, and I had never planned for the 'alt fems' research to become longitudinal or a cohort study. But when I was invited to write this chapter, I realized that enough time had passed (finally) for the women, their lives and their experiences to have moved on. I also realized that I very much wanted to know whether they had finally abandoned their alternative identity and decided to age—as some might say—gracefully. This chapter focuses on how the participants compared their current appearance with their appearance when younger—and how this made them feel, and if they still felt like alternative women. Although they were the most forthcoming about their appearance, they also talked about how they fitted it around families and jobs.

Revisiting the Research

For this chapter, I re-interviewed some of the original participants to hear their updated thoughts on ageing; this also necessitated revisiting the original data to compare opinions or predictions made then by the women, and how (or whether) these had changed. Tracing participants proved to be time consuming and yet more straightforward than I had anticipated, mainly thanks to the Internet but also because I had kept all the information from the first interviews. In total, I traced eight of the twenty participants (over a third), as shown in Table 9.1. Their current average age now is around 48.

Of the eight participants, I interviewed five in person and three online. My first question was 'Can you give me a potted version of the last decade?' The five face-to-face interviews all took place at the participants' homes, and lasted about an hour. There are, of course, differences between face-to-face interviews and those conducted online, as others have discussed (e.g. see Mann and Stewart 2001; Hine 2005; Holland 2008). For this reason, I arranged to telephone the three online participants first, to discuss ethical issues and give them an idea of what we would be talking about before we actually sat down in front of MSN Messenger. This also gave me an opportunity to ensure that they were comfortable with the idea of typing their replies, and that they had allowed enough time to answer fully, taking into account 'thinking time' and that people type at different speeds. The online interviews took slightly longer than the face-to-face interviews, lasting around one-and-a-half hours.

In the original interviews, my own status as an alternative woman helped me to gain access; it also helped with snowballing, and with 'insider' knowledge about shops, clubs, people, bands and so on. My definition of 'alternative' was 'not traditionally feminine' in appearance. In this recent set of interviews, the common experiences were fewer as all of us had less subcultural capital. Thornton (1995: 4) describes subcultural capital as 'a distinction between the authentic and the inauthentic'; she

Table 9.1. Original participants traced

Participant pseudonym	*Age then*	*Age now*
Bee	38	**51**
Claudia	37	**49**
Delilah	36	**48**
Edie	39	**50**
Gemini	48	**60**
Lara	31	**43**
Sparkle	27	**40**
Vash	37	**50**

explains that she sees '"hipness" as a form of subcultural capital . . . [which] confers status on the owner in the eyes of the relevant beholder . . . [It] can be objectified or embodied . . . [and] is embodied in the form of "being in the know"' (1995: 11). The life experiences of the participants were also much more disparate. For example, four of the respondents had retained the same jobs; three had given birth to children; four had married; and two had moved from the United Kingdom to Canada and the United States respectively. Two had undergone major surgery; one had gone bankrupt; and—perhaps unsurprisingly, although in opposition to the 'toning down' discourse—six had had more piercings and/or tattoos. Of course, they had also changed physically, and I was very interested to see whether, on meeting them, in the first moments I would still consider them to be alternative in appearance; in fact, most of them were significantly less alternative in appearance in that there was simply less evidence overall of hair dyeing and alternative accoutrements (such as the big goth or army boots that many had favoured previously). Black clothes were still prevalent as before (Holland 2004: 75), as was the stated explanation of the need for comfortable clothing (2004: 76). Footwear, as in the first interviews, was the subject of much discussion, primarily around the difficulties in finding comfortable shoes or boots that weren't too 'girly' and were also age-appropriate. However, it was still obvious that these were women with a background of being 'different' and outside of mainstream fashions. Their comments on my appearance were also illuminating in that the comments revealed that the women were still aware of the appearance of other women, particularly alternative women. For the original interviews, I had had bright orange and yellow hair with red extensions, and wore black leggings and paraboots. I now had bright red hair and wore black jeans and trainers. I received comments such as 'Good to see you still have freaky coloured hair', 'Good on you, you haven't given up the ghost' and 'Still in black then'. All eight asked me how old I was now. I am not entirely certain that I passed the test—if there was one—but my own toning down was about comparable to theirs.

Changes

There are two main societal changes that have developed since the original research, which have impacted on the participants as women, as ageing women and as alternative women. These are the sexualization of culture and the ubiquity of the Internet. As with the earlier interviews, I found that while there is ambivalence about, and even fear of, ageing, there is also pleasure; there are pressures and tensions, but alongside that is a determination to find a balance between ageing 'gracefully' and remaining true to themselves. The ambivalence stems mainly from the pressures of work or family, which make maintaining an alternative appearance difficult or simply unfeasible. Several mentioned that their children policed their appearance far more than their partners or families ever had—and they were willing to tone down for their children in ways they had never been willing to do for other people. This was an attitude echoed by both Claudia and Diz in the original study: Claudia had said she used to dress conventionally when her daughter was small, in order to be taken seriously by the school and other parents, and Diz said that she and her husband planned to have children, and when they did she would start to dress conventionally for those same reasons.

Additionally, one of the main tensions the participants experienced was still in policing themselves and other women in order to judge how far to go. They wanted to encourage and applaud other women's appearances, especially as they aged, but at the same time found that other women still served to show how far they could go—what was acceptable. The sexualization of culture (McNair 2002; Attwood 2009), where Ann Summers shops are on the high street and burlesque classes are offered in every town, both helps and hinders the ageing alternative woman. On one hand, the pressures are around the images of ageing women in the media, especially celebrity women who undergo secret surgical and non-surgical procedures in order to look younger. But on the other hand, the fact that women like Madonna and Kylie Minogue are in their forties and fifties and are still seen to be strong, independent and attractive (albeit with 'help') does inspire my participants to see their age more positively. Both Amy Wilkins (2004) and Dunja Brill (2008) have argued that being a goth brings feelings of empowerment to their participants. Similarly, my participants—who may have begun their subcultural lives as goths or punks but later became more generically alternative—also said that retaining some semblance of their alternative appearance (and therefore mindset) was crucial to their sense of self and their self-confidence. Unfortunately, alternative female role models remain few and far between, which was also a theme in the original study. The established names—such as Siouxsie Sioux and Patti Smith—remain, but few new comparable female artists are emerging. In fact, several weeks after my interview with her, Vash rang me to say that Ari Up, lead singer of The Slits, had died. 'She was only 48,' she said. 'Younger than me.'

The second societal shift is the huge, and relatively recent, growth of the Internet. The Internet is a useful and welcome tool for them in that they can get in touch with old friends, share photos and memories of the past, and feel some kind of connection with their younger selves—the selves who had abundant subcultural capital, before they began to tone down. In the original study, only a quarter of the participants (five women) mentioned or alluded to using the Internet. In contrast, all eight participants in the current study said that they used the Internet daily, and regularly used social networking sites to get in touch with old friends. 'Mediated interactions have come to the fore as key ways in which social practices are defined and experienced.' (Hine 2005: 1) As Hodkinson explains, subcultural capital is useful as it indicates levels of status within and outside of subcultural groups (2002: 81), and the Internet serves to remind them of the capital they once had. For example, Lara said, 'I love to see pictures, photos, of myself with my friends when I was, like, 20. Photos I might never have seen before. It reminds me what it's all about.' Similarly, Sparkle said:

> I feel I am [sic] part of something. I don't go to clubs or anything any more so you lose touch, you feel, lost, you know? But talking to people online, I use some forums, and I am now in contact with some old friends of mine, it brings it all back. I feel as if it were just yesterday. My appearance was pretty extreme. I just loved myself then.

Such interactions serve to reinforce their feelings of being successfully alternative, of having subcultural capital—both then and, in a more limited way, now.

People (that is, strangers) in general were reported as being equally or more of a point of tension for the participants; in the original interviews, many participants had reported that people had felt able to comment on their appearance (Kiki, Jodie, Morgan, Zeb, Vash, Miss Pink) or even to touch their tattoos (Claudia), to which they responded in a variety of ways from 'stony silence' to running away. Now, though, they find that people—including children—feel able to comment on their appearance quite openly, asking why they look like that ('Why is your hair that colour?' 'Why does that lady have so many tattoos?') and that the parents of the children who ask such questions do not scold their children for being rude. Claudia, who still works in a tattooing and piercing studio, said, 'I say, "Well why do *you* look like *that*?" and they laugh, as if my question is just, you know, not valid.' Claudia echoed the comments she had made in her original interview, where she said she was 'protected' from the outside world because she worked in a tattoo studio, and worked out in a traditional gym where her tattoos were not an issue. Of the eight women I interviewed, five said they felt that their appearance was becoming a double-edged sword: it helped them to feel and look younger than they actually were, but the side effect was that it somehow infantilized them so that people felt able to be confrontational with them about their appearance, rather than treating them as mature adults. However, two of the women also admitted that this had been the case ten years ago, and Delilah acknowledged that women are routinely infantilized and objectified throughout their adult lives.

The Sum of its Parts

Where were points of tension among being 'older', being female and being alternative? The original data demonstrated that, for the participants and women like them, a constant careful juggling act had to be practised between being feminine enough and being alternative enough, and that various compromises or limits were undertaken or imposed by the participants on themselves. Added to those are the compromises or restrictions needed for family or employers. Age brings new issues that make the balancing act more precarious. For example, several participants mentioned weight gain and how it was increasingly difficult to keep fit and lose weight. With Bee, Vash and Gemini, we see that the participants are at pains to assure me that they are still 'themselves' and that this is directly mirrored through their efforts not to tone down too much. This happened in each of the eight interviews. Bee had moved back to her native Canada from the United Kingdom, was no longer a performance artist (instead she is now an administrator for a theatre company), and had put on weight: 'I have gained, oh maybe a hundred pounds. I am BIG.' I asked how she felt about that, considering she had shared her anxieties about her weight in our previous interview (Holland 2004: 12). She replied:

> Well, I see it as part of ageing. You know, it's true what they say, everything gets harder after 40 [smiley emoticon],[1] everything, from walking upstairs to losing that extra doughnut you ate—or the extra four, ha ha. The funny thing Sam is that I still feel OK most of the time. I mean, I know I am not, I waddle I get breathless, I get hot and bothered, I feel that people look at me and think 'she is out of control, she is not clean'. I know I am unattractive to women now. But my hair is still purple, I got some facial piercings when I moved back here, and some funky retro eye glasses. I am—well yes, I am obese. But I am still myself.

Bee's career change was mainly because of her weight gain; she said she no longer felt she had the confidence to perform on stage: 'I perform all the time anyway, kids stare at me now!' In some ways, we see that the fear of being the 'mad cat lady' (2004: 119) was catching up with Bee (I will return to this point again in the conclusion). Gemini also reported weight gain and talked about how it not only impacted on her tattoos, but also on her job and her relationships:

> I go up and down. My tattoos change all the time! Sometimes they are wider than others. And I have always had this relation with myself, where I think I could do better, I could look beautiful, I don't have to be the plain person. But for me, the tattoos and the piercings, they really off-set those feelings. Because tattoos and body piercings, I find them beautiful so by extension, I am beautiful too—or my body is. Except my body is old, I am an old woman. I don't feel it, I don't want to be, but there isn't much point in denying it. But I don't care. I suppose I don't care that my skin is no longer firm, that I am losing my memory and my sense of smell and my sense of taste, and my hearing will fade and my eyesight . . . god, it's depressing isn't it. But I will always have my tattoos, you see, I will always have these reminders of who I am, of what I am or of what I wanted to be.

Gemini's description of being 60 seems exaggerated, imagining a loss of her senses akin to that of a woman 30 years older. Gemini told me that her life at present was not happy because she was single and lonely, and felt that her appearance repelled potential partners; she also felt isolated in her job: 'All they talk about is celebrity magazines; I don't have any interest in that shit. And they see me as a circus freak because of my tattoos and I don't eat meat, you know, they are [long pause] not like-minded.' Gemini and Bee were the two participants who complained most bitterly about their jobs and how they felt that their appearance and 'alternative' approach set them apart from their colleagues. They also both talked about their weight more than anyone else.

In her original interview, Gemini had listed for me the many surgical procedures she would have been willing to undertake to 'improve' her body. One of Gemini's main concerns seemed to be that her weight gain not distort her tattoos. She said she still felt rebellious but that she recognized that 'I am the crone now, the crazy old lady', and this tempers how rebellious she felt she can outwardly be. 'My age makes me feel more vulnerable, physically I mean, I don't want a bunch of hoodies giving me a hard time just because I am the tattooed granny.' This seemed to be the crux of the matter: all the discussion about toning down, whether purposefully or not, reached this point. It is the realization that one day they will be rather frail old ladies, or they will become the 'mad cat lady' they so feared, and they are not sure whether they will be able to deal with the negative attention that this may attract.

Vash previously said she had not 'let herself go' so was 'still here', as if changing her appearance would be the equivalent of disappearing. At the time of her last interview, Vash had gone through a number of challenging years; this time she was very happy and settled, and had worked as a civil servant for many years, being promoted to manager: 'I wear smart-ish clothes, nothing fancy, comfort is still the most important thing.' She again spoke about how she was 'still here', in similarly stoic terms:

> My weight goes up and down but I am still hanging on, I haven't turned townie yet, I will never be mainstream I don't think—I hope! I got married last year and that really helps, that he has always been the same as me, we like the same music, and yes we are getting older, yes I am 50, but I am still hanging on. My husband, I have known him as a friend for years and years, and he loves me as I am. I like to think I will never take my facial piercings out, even though all the little posh student girls have them now. I will always look like this, or an older, more tired version of this. I am happy with that.

Vash seemed to accept that her appearance may tone down and she may seem older or 'more tired', but was philosophical because she was happy generally. Vash's reference to 'that really helps' meant her husband accepted her as she was; she didn't need to worry about trying to change her tastes or appearance for her husband as his tastes were the same. Vash got married on Halloween and wore a black dress:

> My mother wasn't happy but my mother is never happy about how I look . . . We wanted to get married in a way that showed who we were, and who we had been, you know, a bit goth-y and different.

Two other participants mentioned their mothers' disapproval of their appearance:

> My mother is relieved that I have finally toned down my appearance; she would like me to go further. I tell her I have gone further than I wanted to (Sparkle).
>
> My mum . . . oh well, you know what a pain she has always been, very controlling [Lara talked about her mother at some length in her original interview]. She moans on and on about me keeping my piercings, why don't I take them out, when will I grow up. I tell her to shut up, it's not up to her (Lara).

The responses of mothers had been mentioned in most of the previous interviews and, despite the participants all now being over 40, mothers were still cited as an area of tension around both appearance and lifestyle. Much has been written about the problematic mother–daughter relationship; feminist scholarship has argued that mothers exert particular influence over their daughters, and are one of the primary ways that girls are socialized into gendered roles (Chodorow 1978; Ogden and Steward 2000; Usmiani and Daniluk 1997). Ray (2002) found that mother–daughter relationships were often fraught with conflict, partly because of the child's need to become independent (2002: 114), and it appears that this conflict continues well into adulthood, becoming the norm in many cases. Of course, alternative women probably also dress very differently from their mothers, and life choices such as extensive tattooing cause friction in the relationship. Mothers are reported by the respondents to be relieved when their 'alternative' daughter begins to tone down her appearance.

Still Toning Down?

'Toning down' was again, as with other studies (e.g. Bennett 2006; Davis 2006), a key issue. The previous research revealed that

> a common assertion amongst the interviewees was that, although they may have toned down their appearance, they were still the same ('alternative', 'different') person they had ever been . . . In this context, 'toning down' functions as a metaphor for general ageing. The participants consistently refused to say 'I am ageing' and instead replaced this with the term toning down. (Holland 2004: 127)

While this remained partially true, in that the participants insisted they remained 'themselves', they now more readily—if wearily—admitted that they were ageing. As we saw above, family members impacted on the participants' choices about how much to tone down their appearance; even colleagues affected how they felt about themselves (particularly for Bee and Gemini). Lara and Delilah previously had been the most passionate when talking about how they had or hadn't compromised their appearance, and this hadn't changed for either of them. In our last interview, Delilah was a self-employed artist who made cards and sold them to individual shops but has since expanded to selling her cards wholesale. This meant that she worked in a rented studio and was still able to avoid 'being put in a crap old blouse', as she succinctly described it (2004: 91); she worked alone in her studio so could dress how she liked, and this has been a lifelong choice for her. She—like all the participants I re-interviewed—mentioned her last interview and said her anxieties then were nothing compared to those she had now:

> I re-read our last meeting before you came and oh my god, it was bloody hilarious, I mean really. I was stressing away about being 36! Thirty-six for fuck's sake! 36 is nothing, I can't believe I was scared of being 40. And now I am 48. Nearly 50. And I am still trying to find ways to stay myself, to wear what I want, not to compromise for anyone.

I asked whether she thought that, if I were to return again in ten years, she would say the same, that 48 wasn't really that old. 'No, because, you know, because 48 is nearly 50 and 50 is scary.' And yet previously 40 had been scary. Delilah looked remarkably similar in both age and in appearance to the way she did at our last meeting. She voiced a fear that was mentioned in both the original and recent interviews (and which I myself also share):

> I know I look good for my age, and I think I do manage to stay myself, to be how I want to be. It takes a lot of effort you know. I have lived alone for years now, I don't have kids, I work for myself. It is my life's work, if you like. I just don't want to be mutton dressed as lamb. I don't want someone to see me from behind and think I am some girl and I turn round and they see I am nowhere near being a girl. I don't want to see the shock on their faces. I have to keep surveillance on myself; on what I can wear.

Arguably, this is the same for all women—alternative or not—but especially those women who have always invested a lot of time and care in their appearance. Lara agreed that she had toned down her appearance since the birth of her children, but said she had, like Delilah in her last interview, both toned down one aspect of her appearance (not dyeing her hair anymore and wearing what she called 'plainer' clothes) while she added to another aspect (Lara was always particularly keen on her piercings):

> I still keep my work and my own clothes separate. I mean, they are all my own clothes, yes, but I couldn't wear my shabby old clothes to work really. I just wear black all the time. And especially now, I just have less time to think about it all because of my sons. Having children has really—it was a shock, I am sure everyone says that but you know, they take up so much time and energy, every thought ends up being about them. I work part-time now. I read your book again for today and the things I said! But I must say one thing, you know, with my piercings, I have to keep my piercings, they are part of me, that is absolutely the same, totally.

Lara's and Delilah's responses differ in that Lara seemed to be no longer concerned about maintaining an alternative appearance as such. Instead, she was concerned to remain herself, even during what she called the 'maelstrom' of having two small children. She said that spending as much time on her clothes and her appearance as before was simply impossible. Much has been written about how children and family can limit women's access to leisure time and activities (e.g. see Thomsson 1999; Gilroy 1999; Holland 2009; Thébaud 2010), but keeping her piercings did not require time and effort like other aspects of her appearance. I asked Lara, 'If you took out your facial piercings, would you no longer be yourself?' and she replied immediately, 'I would feel very wrong, very bare, very, um, very exposed.' Delilah, with no children or partner, was more able to focus on the maintenance of her appearance.

In the earlier interviews, many of the participants (such as Gwendolin and Kiki) acknowledged that their appearance functioned as a 'protective layer' against negative attention and that, though they were often under-confident, their appearance somehow conversely gave them more confidence. However, several of the women pointed out that they were finding the negative aspects of ageing as an alternative woman to be more worrying than they had anticipated:

> The thing is, I have always been, you know, I have always had men fancy me, alternative men, I have always had a lot of attention. I looked good. And, before [when younger], my appearance, well, I didn't get townie men, or idiots, finding me attractive so they pretty much ignored me. Or if they gave me schtick [insults] I would just ignore them, my appearance was like a mask, or a suit of armour. Now it feels much more like I get stared at for all the wrong reasons (Delilah).

> I was an It girl. I was gorgeous, I was like a peacock, I had blue hair and ten tattoos. Me and my friends, we were popular and gorgeous and we could pick who we wanted. I just ignored the other men, the mainstream men. If they asked me out, I ignored them. If they made some stupid comment on my hair or my outfit, I ignored them. My boyfriends were gorgeous boys in make-up and they had dyed hair. Now, well, now the people I know are the mainstream people, the people I used to ignore. That's who I have to mix with. But I still see myself as the girl with blue hair. So it is kind of hard to fit in, anywhere anymore, not even with the gorgeous kids because I am nowhere near their age (Sparkle).

It would be interesting to be able to pinpoint whether these feelings were truly based on the fact that other people (mainly men) now found their appearance ridiculous, of whether they were more aware of being outside the norm of feminine appearance. Certainly participants such as Sparkle sound adrift in a life where subcultural capital has been diminished by age, and where everyone she encountered no longer understood her appearance or her true identity as an alternative woman.

Edie still owned an alternative clothing shop, and said that she thought working there helped to keep her feeling alternative:

> I went through a little phase, really because—I don't know—I just went through a little phase of not being myself, I was a bit drab, frumpy, you know, I thought I should be and then I thought 'hang on, this isn't me', so I dyed my hair again and now I think I have it

> right. You know, I see all these very young gothy girls come in to my shop and they get everything, it's all there, they don't have to tramp around charity shops and secondhand markets like we did [laughter] and it's just there, the outfit, the—er, the whole package, like they are buying themselves an . . . [*SH*: an identity?] Yes, exactly. It's all prepackaged for them. And I think it's so different, so much less creative, so much less hard work. They make me feel old.

So Edie's daily contact with young alternative women made her more aware of herself as an ageing woman, but also maintained her connection with that world. Since our last meeting, when she had no piercings or tattoos, she has acquired two facial piercings and two large tattoos on her upper arms:

> I got these done because I had always wanted them and I used to have a Mohican, and dreads, and all that, and suddenly I have this short bleached Marilyn hair and, you know, it's not really enough to show how, what, I am, you know? So I thought I would go further. It makes me feel better when the little goth girls come in [to the shop], less like an interloper in my own shop [laughter].

This echoes Flong's comment in the original interviews that 'Joe Bloggs in the street' has body modifications so 'real' alternative people have to have even more (Holland 2004: 157). But Edie also mentioned the same problem that Gemini had: her body was ageing and therefore changing. 'I have bingo wings[2] even though I go to the gym; they just seem harder to get rid of than they used to.' Her tattoos, then, signal her identity as an alternative woman, but also draw attention to the reality of herself as an ageing woman. This is the type of conundrum that became apparent in all eight interviews.

Conclusions

This study has become inadvertent longitudinal research, which implies that I will interview the participants again in a decade or thereabouts. It would, of course, be fascinating to return to talk to these participants yet again, about alternative women and ageing, and all of them asked me if I would and expressed willingness and enthusiasm to be involved. Interestingly, the participants were focused on talking to me about their appearance—how it had changed or not, and how they were managing the feminine/alternative juggling act that they had begun years before. My questions about how they managed jobs, families and other 'grown-up' issues seemed of less importance to them overall. I would posit that this is because they remembered the initial interview as an outlet for an otherwise limited discussion about their appearance and its meanings.

Do the participants still consider themselves to be in some way alternative? All eight said yes. But I also asked all the participants whether they found it more difficult to maintain their alternative identities as they aged, and the answer was also yes. The participants mostly exhibited anxieties about reaching a certain milestone age—predictably ages with a zero on the end were the most feared. Gender still plays a key part in prescribing behaviour, even though these women had rebelled against what was expected of them for most of their adult lives. The participants had spent their adult lives choosing to attract attention to their appearance, but paradoxically using their appearance to deflect negative attention. If we factor in that the participants had toned down their earlier, more extreme appearance, and add to that the increasing invisibility of women as they age (because they are seen to be less sexually available), the experience of ageing becomes doubly disconcerting for them. Their tattoos no longer signalled that they were alternative women, but instead drew attention to their ageing bodies and their struggles with weight gain or other changes. Their appearance no longer served to deflect negative attention, but instead attracted it. Positive attention to their appearance also diminished as they grew older—for example, the long blue hair that once seemed so stylish now framed a 'more tired' face and became incongruous. All eight stated or inferred that they feared becoming grotesque: for example, 'a joke' (Claudia), a 'shock' (Delilah) and 'ugly' (both Lara and Sparkle). Crucially, the fear of becoming the 'mad cat lady', which was so prevalent in the original interviews, was becoming more of a possibility for some of the participants.

Of the participants I re-interviewed, three had created jobs for themselves in which they could continue to dress as they wished: Delilah (artist), Edie (shop) and Claudia (body piercer). Lara had toned down her appearance as a result of becoming a mother and having part-time work; Sparkle had, against her will, toned down her appearance because of her job and where she now lived; and Vash managed to compromise only enough to remain comfortable because her place of work was relatively casual. Bee and Gemini both felt very uncomfortable and misunderstood at work. As I mentioned above, women's leisure lives become more restricted as they age, particularly because of heterosexual relationships, jobs and children, and these things undoubtedly impacted on decisions made by the participants about how 'alternative' they felt able to remain.

Part IV
Ageing Communities

–10–

The Collective Ageing of a Goth Festival

Paul Hodkinson

> *Jo [40]*: I think, because people grow up at the festival as well, we have all grown up together. I have spoken to so many people this weekend and it's strange really, I mean I turned 40 this year and I have spoken to so many people this year—over the last three days—who are either 39 now or who have turned 40 this year . . . the main thing is it really is just a number and OK you may look different in the mirror . . . you may not be able to last until four o'clock in the morning, but you are still out there and a part of the scene and there is still a camaraderie and you are still a part of something.

Introduction

Some of the work featured in this book focuses on longstanding participants who form a significant but smallish older minority within music and/or style communities that remain somewhat dominated by younger people. In these examples, we might argue that the scenes or communities themselves remain fairly clearly within the category of youth cultures in the traditional sense. In spite of their overall longevity, the membership of such groups has a substantial age-related turnover, through the falling off of many participants during their twenties and their replacement by waves of younger recruits. As many studies within this volume and elsewhere have shown, the continuing participation of a minority in their thirties, forties and beyond within such longstanding youth cultures provokes important research questions and conclusions.

As the first of three chapters on the subject of ageing scenes, however, this chapter considers a related but distinct scenario, one in which whole scenes or subcultures gradually become older. In this situation, 'continuing scenes', as Smith puts it, remain—at least to an extent—populated by 'the same body of continuing participants' (2009: 428). Here the tendency for participants to fall away during their twenties is less marked, with substantial numbers remaining involved well into adulthood and towards middle-age. In some cases, this may combine with a reduction or arrest in the recruitment of new teenage participants. Rather than finding themselves in a small minority within primarily adolescent cultures, an adult critical mass of participants may find themselves growing up together.

This chapter explores a case study of such a scenario in the form of the goth scene and, more specifically, a particular twice-yearly goth festival that has been taking place in the seaside town of Whitby[1] in the north-east of the United Kingdom for approximately fifteen years. The goth scene emerged in the early 1980s and has, during its three decades, been centred consistently on distinctive and recognizable forms of dark, macabre music and fashion—most obviously in the form of black hair and clothing. Notwithstanding apparently similar yet somewhat separate recent adolescent developments such as emo, the established goth scene has undergone a substantial increase in its average age, especially since the late 1990s, when I first conducted research on the subculture (Hodkinson 2002).

This broader change has manifested itself in a particularly concentrated fashion at the Whitby Gothic Weekend, a festival oriented to the scene that tends to appeal particularly to older and longer-term participants. Drawing on participant observation and interview research carried out at the festival in 2010, I briefly explore participants' changing experiences of the event as they have become older. I show how developments in the personal life trajectories of individuals were closely connected to a broader evolvement in the feel, ambience and character of the festival itself. And while many such changes were somewhat informal, some had become institutionalized, not least in the format and content of organized events and activities and the nature of the clothes, accessories and other consumables being sold by subcultural retailers.

The Whitby Gothic Weekend: An Ageing Milieu

The Whitby Gothic Weekend (WGW) began in 1994 and developed into a twice-annual festival attended on each occasion by over a thousand goths and centred on live music, DJs and stalls selling specialist clothes, accessories and music. In October 1997, the majority of participants at the festival were below the age of 26 and only a small minority were over 30 (Hodkinson 2002). Over a decade later, the event was dominated by over-thirties and an increasingly sizeable minority of participants were over 40. While goths in their teens or early twenties could sometimes be identified among the crowd, the extent to which they stood out as a result of their age in this particularly concentrated older goth environment was striking. Meanwhile, most of the older participants had been involved in the goth scene since their teenage years or early twenties. In many cases, they had been attending WGW either once or twice a year for several years. Many also had come to know a significant number of the other long-term participants quite well, even though they often lived in different parts of the country and rarely saw one another except at the festival. Seeing these occasional friends and acquaintances—and indeed the larger number of other long-term participants whose regular presence made them a familiar and recognizable part of the event—had become an important motivation for continuing to attend. Every six months, or every year, these ageing festival-goers catch up with one another again—and each time they do so they are all a little older. Over the years, they have grown significantly older together, not only in their years but their appearance, behaviour, priorities, lifestyle and outlook.

Needless to say, the development of individual lives is in some respects both complex and diverse. As I have shown elsewhere, trajectories in terms of work, living arrangements and children vary from individual to individual, as do the ways in which such developments affect or are affected by different levels of continuing involvement in the goth scene itself (Hodkinson 2011). Equally, rather than being a simple, linear process of becoming more 'adult', the progression of years often has been punctuated by what Du Bois-Reymond (2009) refers to as 'yo-yo transitions', whereby changing circumstances may at different points prompt movement both away from and back toward approaches to life understood as more youthful. For example, ruptures such as relationship breakups often prompt a reversal of previous reductions in the intensity of going out, dressing up, drinking and so on. Such complexities and diversities certainly were evident among the Whitby crowd, which encompassed the recently separated as well as the just married, the recently unemployed among the just promoted, and the deliberately childless beside the proud parent.

Yet, in spite of such individual fluctuations, there remained an overall collective direction of travel towards priorities and orientations that tend more commonly to be associated with adulthood than youth. Work, for example, had often acquired greater significance to identity and a greater priority within everyday life, while long-term cohabitation or marriage had become the norm rather than the exception, and children were becoming increasingly common. Alongside physical signs of ageing bodies and an increasing consciousness of being older, the overall tendency toward such developing priorities had gradually transformed not only individual lives and friendships but also the overall collective feel of WGW itself. Of course, not everything had changed. The festival still centred on dressing up, drinking, inebriated socializing, dancing, watching bands, staying up late and so on—some aspects of which may well be understood as quintessential 'youthful' forms of activity in one form or another. Indeed, the status of the festival as an occasional large-scale event meant that individuals often felt more able to 'let themselves go' than would normally be the case in their daily and weekly lives back home. However, as I show below, elements of the intensity and character of the WGW milieu had nevertheless evolved in various respects.

Socializing and Conversation

Socializing, catching up with friends and meeting new people have always been at the centre of the WGW experience, as is the case with a range of other festivals and conventions within the world of music and culture (Dowd, Liddle and Nelson 2004).

However, the length of time that many people had known one another and the extent to which their six-monthly reunions had become an ingrained routine were striking. Equally importantly, on wandering round the clubs, pubs, cafes and streets of Whitby over a decade since my original PhD research on the festival, I could not fail to notice the change in the subject of people's conversation. Clothing and music had remained important, as had discussion of different events taking place at the festival and the escapades of the night before. Yet it was also common to observe or participate in conversations about weddings, property, mortgages or careers, which displayed what many would regard as fairly typical traits of (mostly) middle-class adulthood. Furthermore, many were conscious of such changes and of the irony, as they saw it, of sitting at a goth festival, listening to goth music and wearing goth clothes while discussing their respective career woes. The following field note describes a conversation I had with a long-term festival attendee, who noted the number of people who had established successful professional careers:

> *Field note, April 2010*: As has become standard at Whitby, a conversation with an old acquaintance of mine from Birmingham begins on how our respective careers are going. As part of this, I tell him about my research. He observes that one of the most striking things he has noticed in relation to the subject is that way that, when sitting round a table at Whitby with a group of goths, there is a stark contrast from a past situation when most were students or in casual work or temping—and when these things barely were mentioned in conversation. In contrast, he now finds himself surrounded by teachers, accountants, IT consultants and so on and discussion of these aspects of people's lives has become a key part of catching up with them.

Changes in people's identity, status and priorities in the world outside the goth scene, then, did have an impact on their interactions within it, even at an event which, for many of them, acted as a temporary reprieve from certain elements of this world.

Embracing Parenthood

A further subject of conversation that had become more common among goths at the festival was children. Not surprisingly, this was particularly the case for those who already were parents, but a collective awareness of greater and greater numbers of people having children meant that it had become a topic of broader interest. In fact, more than one person spoke semi-disparagingly of a 'goth baby boom', as though having a baby had become some sort of internal trend within the subculture and, more specifically, the festival itself. As one respondent explained to me, the small initial minority of goths who were parents up to a few years ago had expanded substantially in recent years. And this expansion seemed to connect clearly with the fact that significant numbers of goths were reaching their mid- to late thirties, a point after which female fertility tends to drop substantially. She contrasted this apparent collective conversion to the idea of parenthood with what she perceived as a more youthful, carefree attitude to life in the past:

> *Jo [40]*: I'm finding—I don't know whether it's the fact that goths want to hang on to their youth as long as possible or whether they are all of a mindset that they don't want kids, they are the eternally young. And then their body clock starts going tick, tick, tick and 'let's do it now!', but . . . a lot of people that have come to every Whitby are suddenly turning up with little ones . . . there is a lot of people starting to turn up with children now that I have known for a long time and there is certainly a more family environment to it.

Sure enough, during the daytime at Whitby—and also sometimes at the evening events—it had become usual to see goths in their thirties or forties carrying babies, or walking hand in hand with young children. Often, though not always, the infants themselves were dressed in some sort of goth-oriented clothing for the occasion. As with some of the Northern Soul parents described by Nicola Smith in Chapter 12, the bringing of children to events such as Whitby had become commonplace, and as a consequence the festival had started to acquire an increasingly family-oriented atmosphere. Indeed, such was the enthusiasm of new parents to bring their children to Whitby that the festival staff had found themselves quickly developing policy with respect to the circumstances in which it would be permitted for children or babies to attend different parts of the event, as the following field note illustrates:

> *Field note (April 2010)*: A male goth, perhaps 35, very long dark hair, black clothes, fairly standard non-elaborate black jeans/T-shirt 'trad goth' look, but carrying a baby in a front sling, approaches the main festival info desk and enquires as to whether it is OK to bring his baby to the evening event. He is informed that it would be fine to bring the infant into the foyer area—which is the primary social area at the event—but that young children may not be taken into the main live music performance room unless they are wearing appropriate ear protectors.

The realities of parenthood and of children, then, were strongly contributing to a changing milieu throughout the festival.

Partying Softer

The status of WGW as an occasional special event, which lasted a few days and took place in a seaside town far away from everyday work and domestic routines, meant that many participants felt able to participate with greater intensity, whether in respect to appearance, inebriation, late nights or general behaviour, than they normally would in their goth participation closer to home. This is consistent with findings about festivals elsewhere. Bennett (2006), for example, notes the importance of concentrated participation at festivals such as Punk Aid, even for older punks whose involvement had at most other times become less intense. Bennett draws here upon earlier work by Dowd, Liddle and Nelson (2004), who point out that 'the intensity of a festival compensates for its infrequency' and that 'fans and performers can immerse themselves' because participants are 'drawn together from geographically dispersed locations and away from the expectations of everyday life, fans and performers can immerse themselves' (2004: 149). Similarly, Bengry-Howell and Morey (2010) emphasize the importance of music festivals as an intensive collective removal from an everyday neo-liberal world dominated by work and individualism.

In spite of this concentrated festival atmosphere, however, many people's participation at WGW had become a little lower in intensity or extremity than had been the case in the past, and this too had a cumulative impact on the event's collective feel. Although drinking remained an important feature of the weekend, the amounts of alcohol consumed were often at least a little lower—and while previously a substantial minority of participants took drugs such as speed at some point during the weekend, my recent research suggested this had become less common. Similarly, while flirting remained widespread and visible, the number of people involved in long-term monogamous relationships meant that 'pulling', or picking up members of the opposite or same sex, was a somewhat less central and visible feature of the evening events than had been the case several years earlier. In contrast to Taylor's account of the enduring sexual transgression in the queer scene (Taylor, this volume), in the case of WGW, an event that once had had a something of a reputation for an atmosphere of sexual promiscuity and a degree of experimentation had apparently undergone a gradual process of normalization, with the majority of its participants having pursued heteronormative approaches to sex and relationships as their adulthood had progressed.

Further evidence of an apparent reduction in collective intensity could be identified through observation of audience behaviour during the main band performances at the weekend. While dancing to DJ sets remained popular, most of those watching live bands had a greater propensity to stand fairly still, at arm's length from one another, and politely foot-tap rather than dance, surge forward, mosh or stage-dive. The field note below refers to the crowd for the headline band on the second night of the April 2010 WGW:

> *Fieldnote (April 2010)*: The crowd are standing, clustered towards the stage, moving closer as the performance begins, but much further apart than in the standing area of many large gig events, where being squeezed up next to nearby people is compulsory. Many stand fairly motionless, while others engage in polite, slightly awkward and understated rhythmic movements. Some—not necessarily at the front—dance more energetically but they are in the minority, as for other performances that evening.

In contrast to the situation described by Bennett (2006), and by both Tsitsos and Gibson in this volume, in which a majority younger crowd moshes towards the front of gigs while older participants stand and watch or socialize towards the back (also see Fonorow 1997), in this case the whole audience had aged to such an extent that their collective behaviour in watching bands appeared to have changed.

Nights out at Whitby also tended to be somewhat shorter than before. In the past, the main event would finish at 2.00 or 3.00 a.m., and substantial numbers of people would stay out longer than this, whether at private parties held in the accommodation of one participant or another, or via the more public option of walking en masse up to the clifftop churchyard[2] in the town and sitting talking and drinking until the early hours. More recently, however, such post-event activities had become less common, with most participants going straight back to their accommodation after leaving the main venue. Meanwhile, although the main venue still remained open until the early hours, the number of people still present at the end had become strikingly low, with large numbers leaving from about midnight onwards. When I discussed this with respondents, a range of reasons were provided, most of which related in some way to ageing. Some, for example, placed emphasis on the increasing physical impact on their bodies of late nights, while others focused on a reduction in the motivation to make the effort to stay out late as a result of being at the festival with a long-term partner. Consistent with this, those who were single often stayed out later than those who were married or in a cohabiting relationship.

For those with children, the situation had become particularly challenging. Most brought their children with them to the festival rather than arranging child care back home. In a minority of cases, babies were even taken to the evening event, which placed obvious limitations on the length of the evening, levels of intoxication and so on. More often, couples would sacrifice the ability to go out together, taking it in turns to go out and take on childcare responsibilities back in the hotel room. Similarly, when one or both partners did get to go out, they tended not to stay out very late because of the responsibility to care for children early the next morning. In fact, for many other parents the experience of Whitby had changed significantly, to the extent that it had become something of a regular family holiday, centred on their children's enjoyment as well as their sleep patterns. A range of daytime seaside activities had acquired as much importance as the gigs, clubs and subcultural markets:

> *Susan (33)*: Whitby is more of a family-oriented thing for us now . . . 'cos obviously we've got [name] to consider. She won't wanna look round the stalls for hours on end and so we'll take her to the beach and that kind of thing . . . And [name] absolutely loves Whitby . . . there's a lot of goths up there with children now. As Whitby has gone on, it's, it's a family thing for a lot more people.

In various respects, then, a festival which once had been dominated by unmistakably youthful forms of excess and inebriation had apparently shifted down a gear or two in terms of its level of intensity.

Gothing Up—or Down?

In the same way that it often encouraged a greater than usual intensity of participation in terms of people's broader approach to going out during the weekend, the occasional festival status of WGW meant that, for many participants, outfits remained striking and elaborate, even for those who rarely had the time or inclination to spend much time getting particularly 'gothed up' at other times of the year. Equally, many of the broad themes of collective goth style on view at Whitby had remained constant, from the overall emphasis on dark, macabre themes to particular familiar styles of makeup, jewellery and so on. In many ways, the continuing involvement of so many older goths in the exhibition of such overtly subcultural appearances is noteworthy in itself as an apparent rejection of at least some of the dominant conventions of mainstream adulthood. But even in this high-intensity festival environment, there were also visible signs of collective change—a good deal of which was attributable to ageing.

Even a cursory initial observation of the goth crowds walking in the town's streets or sitting in its bars revealed a variety of physical signs of ageing, from balding heads to larger midriffs, coarser skin and older-looking faces. Equally visible were the numerous ways in which goths have—both individually and collectively—adapted their choices of clothing, jewellery, makeup and hairstyle to these changing bodies, as well as sometimes to their developing consciousness of themselves as older, adult or sometimes middle-aged. In this respect, goths' strategies of age-negotiation were comparable to some those outlined by other studies, notably Samantha Holland's (2004) emphasis, in her study of older alternative women, on balancing the desire to remain 'alternative' with an anxiety about being seen to age inappropriately (see also Chapter 9 in this volume).

For both male and female goths, there was a greater emphasis than before on clothes that covered up more of their bodies—specifically legs and midriffs—than previously. Looser fitting and more 'comfortable' outfits also were sometimes preferred. Many women I spoke to said they were less likely than in the past to wear miniskirts with stockings or tight, cropped tops, preferring longer dresses, or a smarter 'corporate goth' look, or sometimes the simplicity and comfort of black jeans or combat trousers and a T-shirt. A significant exception was the enduring popularity of corsets, which served to obscure expanding waistlines while enabling the display of breast cleavage. The continuing veneration of this particular look connected to a long-standing interest among some in a partly Victorian-inspired 'trad goth' look, and this had been embraced by many older goths as a way of continuing to exhibit a striking subcultural appearance. A more recent steampunk variation, meanwhile, had combined the emphasis on Victorian corsetry and so on with a variety of technological and other (usually brass or brown coloured) accessories loosely associated with the steam era, from telescopes to cog wheels to goggles.

For men, the emphasis on outfits that were more modest with respect to bodily display was accompanied by a broader collective shift away from the forlorn, androgynous goth look that had often dominated WGW, alongside the goth scene more generally, in its earlier years (Hodkinson 2002; Brill 2008). Though they could certainly still be seen from time to time, feminine forms of clothing and extensive makeup on men had become less prevalent, while more traditionally masculine features—including goatee beards, shorter haircuts and stockier builds—were more widespread. When asked, many said they felt it was less possible to 'get away with' feminine clothes or makeup once bodies had become larger, faces coarser, body hair denser and, in some cases, head hair sparser. Jon, for example, explained that he no longer wore makeup, had reduced the length of his hair and moved towards looser-fitting clothes primarily because of his awareness of changes to his body:

PH: Could you ever imagine yourself dressing up, more like you used to, or is that just gone?

Jon (38): Because of my physical appearance now, I think I wouldn't feel comfortable doing that. I don't think I would look how I wanted to look at the time so I accept that I would perhaps not dress as loudly . . . I'm larger is the biggest thing—the years have not been too kind to me in terms of weight and things like that . . . and I accept that I don't look very good in tight jeans and a frilly shirt. I would tend to feel like I was more in fancy dress than in going out attire, perhaps . . . I think the slow acceptance that I am not the young person that I was . . . there is a period of denial, but there does come a point where you realize you are what you are and you need to temper your expectations towards fashion accordingly.

Although practice remained varied, the overall apparent drift towards forms of gender display which are at least a little more compatible with normative understandings of adulthood contrasts somewhat with the deliberate retention of queer dress among participants in Taylor's contribution to this volume.

More generally, in spite of the one-off intensity generated by WGW as a festival, and the striking manner in which so many had continued to participate through dressing up distinctively and going out late into their thirties and forties, the collective look of those at the event, as well as its broader feel, had gradually altered over the years as a result of the cumulative impact of participants' individual negotiations through their advancing years.

Adapting to an Ageing Clientele: Promoters and Entrepreneurs

In addition to developments in the collective milieu of the festival, such as those I have described, the communal, subcultural nature of the process of ageing in the case of the Whitby Gothic Weekend is particularly clear when one examines the ways in which it has been reflected in, and hence institutionalized by, the content and organization of the festival.

There is an increasingly nostalgic feel to the line-ups of bands booked for the event. The October 2010 event, for example, was headlined by Wayne Hussey and Simon Hinkler, former lead vocalist and guitarist from 1980s and 1990s band The Mission. The bill for the event also featured Manuskript, which had played at the first ever Whitby festival in 1994 and was playing in 2010 as part of the 'Big 20' anniversary celebrations. As part of this performance, repeated ironic references to age and ageing were made by the lead vocalist, and these were pitched to clearly include the ageing audience in the joke. It was not just the clientele who were long-term participants in their thirties and forties, then, but also many of the performers.

This parity between fans and performers was of significance to people's developing approaches to subcultural participation. Notably, some participants explained that, in searching for ways to adapt their appearance to their mounting years, they sometimes looked to certain performers who were deemed to have aged in an effective or appropriate manner:

> *Jon (38)*: I think your aspirations and your models change as well . . . I look at people like, for instance, Rammstein, the ways he [Till Lindermann] has aged, he has aged very gracefully and wears blue jeans now, a shirt which looks very, very cool, and that is the aspiration I have now to do things simply, feel comfortable.

The tendency towards a nostalgic emphasis on a past golden age in the organizational orientation of WGW as an event, meanwhile, was also visible in other elements of the festival's entertainment. The majority of the music played by DJs after the band performances in the main venue, for example, could be dated to the 1980s or 1990s, with only occasional examples of newer tracks. Meanwhile, a new DJ night called Nostalgia had emerged on the Sunday night of the festival, centred even more specifically on goth and related genres from the early 1980s. Its main competitor on this night of the festival was a longer-running night centred on 1980s pop music. This emphasis on nostalgia was evident in some parts of the goth scene across the United Kingdom and beyond, but it was even more concentrated at Whitby, which had become particularly associated with older participants with an enthusiasm for what sometimes was referred to as 'trad goth'.

A broader diversification in the range of events taking place at the festival had also occurred in its most recent few years—whether such events were organized by the main promoter or a diverse range of volunteers and entrepreneurs. Some offered alternative options in terms of familiar themes such as DJ sets and dancing, but others consisted of a range of other activities, some of which seemed to reflect the ageing nature of the WGW clientele. These had included, on different occasions, a cabaret evening, a comedy club, a pool competition, a dog walk and a children's disco. One respondent commented that many dog walkers had taken to adorning their dogs with goth-style hats or frilly material as part of the occasion. The children's disco, meanwhile, had run for the first time in April 2010 and was regarded by the festival organizer as part of a broader transition of WGW towards a more family-oriented event. This also included trying to make it easier for parents to deal with childcare needs, as she explained:

> *Jo (40)*: that's something we need to look at a bit more, as people start to come here with families. What tends to happen is they start off looking everywhere for child minders or a nursery. If you go to Butlin's or Pontin's for instance, they have a crèche that stays open day and night, where you can leave your kids for a few hours and your mum and dad can go out, and I have looked maybe for nurseries and things, and lots of registered child minders to see if they would open, or some of them would like to run some sort of crèche and I just can't get anyone interested . . . the way it works is that mum and dad buy a wrist-band for the weekend, mum comes out one night and dad comes out the other and so they pay £40 each for one night out which is something we need to look at, a family wristband.

Another consideration in her mind had been some sort of a singles or speed-dating event, or alternatively some mechanism whereby single people looking for a relationship or sex could identify one another amongst the (increasingly spoken-for) crowds. While she was at pains to emphasize that WGW still attracted many single people, the idea that they might need or benefit from a dedicated event or means of identification draws attention to their minority status and the way event organizers were looking at ways to adapt to this collective development.

A further example of the adaption to an ageing clientele of the organizational infrastructure of WGW was the way in which retailers at the festival had adapted and developed their stock of goth clothes and accessories. Sometimes, this might involve adapting to the changing clothing needs or wants of older goths by placing greater emphasis on trends or styles of clothes flattering to older bodies. But there was also increasing emphasis on items with a more overt and direct association with adult lives and priorities. Goth-themed clothing for babies and small children, for example, had become particularly visible when walking round the stalls, and seemed to have become a significant business opportunity—something evident from the number of infants around the town wearing such clothes.

Another example was provided by a stall specializing in particularly elaborate forms of goth clothes and accessories oriented to special occasions, whose primary market had become goths getting married or attending weddings. The stallholder explained that the goth wedding market had expanded significantly in recent years as goths had collectively become older and, in her view, more family-oriented:

> *Jackie (43)*: I was just doing it as a hobby, and now that has snowballed and I am doing full goth weddings and everything now . . . I mean there are a lot of people getting married now, it's like . . . from people that were goths when it started . . . they are at the age

> now where they are having families and getting married and settling down now, so it [her business] does seem a lot bigger—and I mean people are coming here [i.e. WGW] to get married a lot now.

In various respects, then, the gradually more adult-oriented feel of the Whitby Gothic Festival milieu had become institutionalized as a result of changes to the commercial and organizational orientation of the event which would, in turn, feed back into the interactions and orientations of participants themselves.

Conclusion

In this chapter, I have used an examination of the changing feel and character of the Whitby Gothic Weekend to illustrate the ways in which the processes of ageing within a music and style community can take a collective form. In many ways, of course, it is striking that so many goths continue so substantively to participate in such a distinctive music- and style-based subculture and to attend events such as WGW as part of this. That, in itself, is of great significance as a collective phenomenon, demonstrating that questions of older subcultural participation are not limited to the consideration of minority adult participants within communities that remain dominated by generation after generation of adolescents. Likewise, the continuing importance to older participants of many of the fundamentals of the subculture from its and their earlier years—including maintaining a distinctively goth appearance, going out, dancing to goth music, drinking and so on—is noteworthy, particularly with respect to the integration of these forms of activity retained from younger years into older lives and reconciling them with adult expectations.

My primary focus here, however, has been on the ways in which participation in WGW has collectively been transformed in coordination with the individual development of older identities, bodies and orientations. And while, on an individual level, such negotiations of ageing encompass a variety of changing trajectories, I have suggested that there remained a reasonably clear overall collective direction of travel towards a festival with a noticeably more adult-oriented feel. Whether in relation to topics of conversation, the adaption of goth fashion to ageing bodies, the increasing presence of long-term monogamous couples and children, or the earlier nights and reduced levels of inebriation, the event has gradually shifted over its fifteen years away from the greater extremes of youthful hedonism and towards a lower-intensity and adult-centred event for long-term participants of a subculture that is ageing collectively in a variety of ways. Those attending the Whitby Gothic Weekend, through collectively developing their approaches to goth participation in coordination with their changing bodies and lives, were—quite literally—growing up together. And for some there was no discernible end to this ongoing collective negotiation between continuing subcultural identity and ageing lives—and no end to the importance of the Whitby Gothic Weekend itself as part of this:

> *Jackie (43)*: I think Whitby is going to be the goth retirement—I am seeing souped up wheelchairs with bats flapping behind them and flames at the wheels, coffin-shaped motor scooters, I can see it happening, I really can and I think Whitby is going to be like a, you know, you get Rhyll and that, Llandudno is an ideal one for old people—I think Whitby is going to be the goth old people's retirement seaside village . . . Little black tea cosies and things round the fireside, listening to Sisters of Mercy!

–11–

'Strong Riot Women' and the Continuity of Feminist Subcultural Participation

Kristen Schilt and Danielle Giffort

Introduction

Subcultures have long been theorized as the domain of the young. Illustrating this connection, studies of the varieties of subcultural participation often use the terms 'subcultures' and 'youth cultures' interchangeably (e.g. see Epstein 1998; Skelton and Valentine 1998). This connection between subcultures and youth stems from a particular framing of adolescence as a transitional time between being embedded in school and families of origins (childhood) and entering the world of employment and marriage (adulthood) (Hall and Jefferson 1976; Frith 1983; Brake 1985). Yet subcultural participation as a 'magical solution' (Hebdige 1979) to the vagaries and stresses of adulthood historically has been a male domain. As Mike Brake notes, 'If subcultures are solutions to collectively experienced problems, then traditionally these have been the problems experienced by young men' (1985: 163). As young women in the 1960s and 1970s experienced greater parental control and social sanctions for taking up space in the streets (McRobbie and Garber 1976), many researchers of the era theorized that girl cultures started and finished in the bedroom (Frith 1981)—and thus afforded them less investigation than the more publicly visible and accessible cultures of boys (Brake 1985; McRobbie 1991).

Subsequent generations of researchers have intervened in these early conceptualizations of subcultures in two ways. First, the participation of girls and women in male-dominated and mixed-gender subcultures, such as punk rock (LeBlanc 1999), goth (Wilkins 2004) and queer (Halberstam 2005; see also Chapter 2 of this volume), has been documented widely, as has the formation of 'girl-centred' subcultures, such as Riot Grrrl, which take as their focal concern issues widely shared by women and girls (Kearney 1998, 2006; Schilt 2003a). Second, the premise that subcultures are only for the young has come under investigation. As Andy Bennett (2006) notes, the assumption that personal connections to music and subcultures should fade in early adulthood ignores the ways in which music is commercially marketed and how fandom can become a core part of a person's identity. Taking the punk subculture as an example, some punks find strategies that allow them to negotiate continued community participation—even after they 'grow up' and 'get real jobs' (Andes 1998; Bennett 2006; Davis 2006). While punks might transcend the punk aesthetic as they age (Andes 1998), they can still feel they embody a punk ideology that gives them a continued connection to a local, translocal and/or virtual scene (Bennett 2006). This focus on ageing subcultures calls into question a view of subcultural participation that is static and time-bounded, emphasizing the dynamic shifts that can occur across participants' lives.

What is missing from this body of research is a merging of these interventions: women's subcultural participation and ageing. With a few exceptions (Halberstam 2005; Vroomen 2004), research on ageing subcultures and fandom has illuminated more about men than it has about women. In this chapter, we seek to add to this literature. We focus on girls' rock camps, week-long summer-camp programs in which adult women teach young girls how to play instruments, form bands and put on shows. Drawing on in-depth interviews and participant observation at girls' rock camps in two major US cities, we show how the organizers and volunteers constitute a useful case study for examining women's strategies for continued subcultural participation. Many of the volunteers and organizers come to rock camp with a history of participation in the Riot Grrrl movement in the 1990s, and continue to be involved as adults in their local music scenes as musicians and fans. Rock camps provide these women with an institutional location in which to share punk feminist ideologies, women's musical history and technical knowledge intergenerationally. In contrast to ageing men in punk scenes, who may seek to establish themselves as authorities on punk history and authenticity (Bennett 2006; Davis 2006), these women strive to form collaborative efforts with young girls and to learn from the younger generations' experiences. With their alternative appearances and often nontraditional career and family lives, the organizers also show dynamic trajectories for both adulthood and subcultural participation.

Girls' Rock Camps in the United States and Abroad

Since the first girls-only rock camp, the Rock 'n' Roll Camp for Girls, opened in Portland, Oregon, in 2001, over two dozen rock 'n' roll camps for girls have popped up all over the world. Each camp is independently run, although all have adopted similar projects of empowering young girls through rock music. The camps share an outlook that girls and women continue to be marginalized as producers in the rock community. The goal of the camps is to make the world of rock more gender integrated and, at the same time, to find positive strategies to increase girls' self-esteem and sense of agency. Consequently, girls' rock camps are not just music camps; these programs are about fostering positive self-esteem and building a community of girls and women who actively resist their cultural subordination and work to promote social change.

Each camp has its own mission statement and camp program, an autonomous structure similar to the Ladyfest conventions that arose across the world in the 2000s (see Schilt and Zobl 2008). The week-long camps offer day programs during the summer for girls between the ages of 8 and 18. Campers learn how to play a rock instrument,[1] form their own rock band and, at the conclusion of camp week, perform their own original song at a local venue. In addition to offering instrumental lessons and band practice sessions, the camps provide workshops. Topics vary by camp, but common sessions include the history of women in rock, screen-printing, self-defence and zine-making. Most rock camps also invite local women in bands to come to perform. The cost of attending varies from camp to camp, but is generally around $350. Most camps, however, accept girls into their program regardless of ability to pay by offering financial aid to cover part or all of the cost of enrolment.

To facilitate camp week, these organizations require a strong and dedicated group of volunteers. Primary volunteer positions during camp week include band coaches, who help guide the girls through band practice sessions; counsellors, who are responsible for looking after a single band and getting it from one activity to the next; and instrument instructors, who teach girls the basics of their respective instruments. Musical knowledge is not required of all volunteers, although most volunteers actively participate in their local music community in a variety of roles—as fans, musicians and behind the scenes as concert promoters, music journalists and tour managers. While organizers and volunteers come from a wide variety of musical backgrounds, they share a dedication to the key concept of rock camps: adult women teaching young girls how to make music. It is this generational diversity in a gender-segregated space that makes rock camps a useful setting for examining how women's subcultural participation in feminist, queer and punk music scenes continues over time.[2]

Background to the Study

The research presented in this chapter is based on in-depth interviews and participant observation at two girls' rock camps: Girls Rock! Midwest (GR!M) and Girls Rock Southwest (GRSW).[3] Both camps are located in cities that have vibrant independent music scenes that are inclusive to feminist and queer punk musicians. Midwest City and Southwest City also have several local universities and thriving retail and technological sectors that are willing to employ people with 'alternative' appearances—such as tattoos, piercings and dyed hair. This chapter's second author observed at GR!M during the summer of 2008 and at GRSW in the summer of 2009, where she participated as a volunteer drum instructor and band coach. Between August 2008 and December 2009, she conducted thirty-two interviews with volunteers from GR!M and GRSW. Participants ranged in age from 21 to 47, with the average age being 32. Almost all participants report playing an instrument, and twenty-two participants are currently in—or have previously been in—a band. Playing music is not a full-time profession for most participants, although a few work in the music industry. Most participants hold day jobs in positions such as administrative assistant and teacher. Many of the volunteers mention directly participating in and being influenced by the music and ethos of the Riot Grrrl movement.[4] Consequently, our discussion in this chapter will also draw on the research of this chapter's first author on the formation of the early Riot Grrrl scene (Schilt 2003a, 2003b; Schilt and Zobl 2008).

From Riot Grrrls to Riot Women: The Roots of Girls' Rock Camps

In the early 1990s, Riot Grrrl attracted a great deal of media attention as a 'spectacular subculture' (Hebdige 1979) organized by and for young women. Reporters typically emphasized the young age of the participants and anticipated a quick decline of Riot Grrrl after members 'grew up' (Spencer 1993: C2) and 'hit . . . the adult real world' (Chideya 1992: 86). Riot Grrrl members, in contrast, charged that while the subculture might dissolve, the feminist ethos and the Do It Yourself (DIY) practices of subcultural production would continue to live on (Juno 1996; Klein 1997). As one member argued, 'Riot Grrrl is not something you grow out of . . . I think we're going to be strong Riot Women.' (Juno 1996: 181) Discussing girls' rock camps with the organizers suggests this promise of continuity has been realized. Rock camps provide an institutional role for women who want to continue their participation in punk feminism in new ways in adulthood. Karen, a counsellor at GRSW, discusses why this kind of role is important to her. She praises Riot Grrrl for focusing attention on girls and girlhood. Yet, as she's gotten older, she's started to question the absence of adult women from the movement, arguing that Riot Grrrl 'should encourage involvement from girls *and women*'. For Karen, her Riot Grrrl identity no longer feels legitimate because of her adult status; however, she is still committed to the ethos of the movement and wants to find a way to stay connected to it. By participating in rock camps, she challenges the idea that she might be 'ageing' out of punk rock feminism, and instead developes new ways to connect women and girls.

For Karen and many of the other organizers, Riot Grrrl is not a static subculture that died out in the 1990s. Rather, opportunities for involvement in and the expression of Riot Grrrl politics have evolved over the years in new formats. Like Ladyfest conventions (see Schilt and Zobl 2008), girls' rock camps have emerged as a way for adult women to continue being punk feminists and musicians beyond their teens. For many women, rock camps could not have happened without Riot Grrrl. These organizers credit Riot Grrrl with being a big part of how they started playing rock music, why they became feminists and why they now participate in girls' rock camps. Tracy, a board member and band coach at GRSW, recalls the first time she saw Bikini Kill—a well-known Riot Grrrl-associated band—in the early 1990s:

> Before that [show], I had seen some women play, but it was mainly guys. Seeing them [Bikini Kill] really inspired me. They were up there playing and not apologizing for it . . . it was like 'Fuck you, I'm going to do this.' That's when I thought, 'Wait a minute, I can do that, too.' It never occurred to me to pick up a guitar until Riot Grrrl happened.

Riot Grrrl challenged the traditional roles of girls and women in the punk scene and mainstream society so that girls like Tracy considered playing the electric guitar—a practice often equated with men and masculinity (Bayton 1998)—to be a possibility. Participating in Riot Grrrl—whether by making zines or playing in a band—taught participants, as Corin, a counselor at GR!M, explains, 'that instead of being an observer, I can be a participant' in the punk scene. Not only did Riot Grrrl encourage girls to 'use guitars as an outlet for self-expression' (Carrie, GR!M) and to 'let us express our feminist politics through music' (Donna, GR!M), but it also offered a space to challenge what many participants saw as restrictive gender expectations for women. Kaia, a board member at GR!M, explains how Riot Grrrl 'changed my own relationship with masculinity and femininity', allowing her to break from traditional feminine appearance expectations. Through their involvement in Riot Grrrl, interviewees viewed music production as a positive outlet for self-expression and a means of political resistance and activism.

Reflecting on the importance of Riot Grrrl in her life, Donna says, 'We had success—we got heard, we took up space, and as a feminist and a mother, those are the things that I still see challenging girls today.' This feeling that there is still important feminist work to be done characterized many of the interviews. In order to address this continuing need for integrating girls and women into music scenes, volunteers who participated in and were inspired by Riot Grrrl politics took that ideology and turned it into girls' rock camps. Kathi, a board member and band coach at GR!M, explicitly links the emergence of girls' rock camps with Riot Grrrl, as she claims, 'Riot Grrrl had to have happened in some way to make rock camp happen.' Janet, a workshop leader at GR!M, adds, 'It [Riot Grrrl] is what put the bug in our ear to do this in the first place.' The general feeling among volunteers is that while Riot Grrrl created a space for women in rock, female participation remains something remarkable or unusual. A focus of the rock camps is to take the grassroots ethos of Riot Grrrl and put it in a more formally organized and reoccurring setting as a way to reach a wider audience and, as a long-term goal, to mainstream the idea of girls and women playing rock music.

Describing the camp program at GRSW, board member Kathleen states, 'Everything about camp is totally Riot Grrrl. Indeed, many volunteers characterize girls' rock camps as Riot Grrrl in action, specifically pointing to the feminist and DIY focus of these programs. Hanna explains how she and other organizers 'came of age in terms of our feminist motivations with Riot Grrrl . . . that's a huge thing influencing us, so it's sort of unavoidable that we bring those ideas into what we do [at the organization]'. Drawing on the DIY ethos of punk rock feminism, girls' rock camps are designed to teach girls how to gain confidence in being a cultural producer. By having campers write their own songs or learn how to use sound equipment, volunteers feel they are helping transform girls from passive consumers to active producers of culture. Similar to the underlying ethos of Riot Grrrl (Kearney 2006; Schilt 2003a), the logic behind these actions is that cultural production is a means towards empowerment and resistance. Giving girls access to creating their own musical and written expression, in the form of bands and zines, provides grassroots strategies for resisting the cultural devaluation of girls and girlhood.

Teaching Punk Rock Feminism Across Generations

Girls' rock camps present a new strategy for continued subcultural participation for adult women—educating new generations of girls about women's music history, feminism and DIY cultural production. Most volunteers connect their motivation for getting involved in these camps with their desire to 'help the next generation of girls' and to stay involved with local music scenes. Karen says, 'I like the idea of passing on and creating a narrative, like making sure more things didn't get lost in terms of feminism and feminist thinking within music history in particular and popular culture in general.' Rock camps allow volunteers, as Janet puts it, to 'actually do something positive to affect the lives of young girls now' and to give girls what Riot Grrrl gave them: the confidence to pick up a guitar or make a zine and discover the possibilities of resistance through cultural production. Such knowledge transfer between generations emerges as a major means through which these adults maintain their involvement with their youthful Riot Grrrl ideologies. Volunteers at girls' rock camps do not claim to be the ultimate authorities on feminism, but rather envision their role with rock campers as 'role models, educators, and cheerleaders' (Becca, GRSW), who facilitate—but do not direct—girls' engagement with cultural production. Volunteers do not, for example, tell the girls, 'This was Riot Grrrl' or 'This is what a feminist is'. In contrast, they attempt to provide a set of thinking tools about Riot Grrrl and feminism, and allow girls to build their own perspectives. This 'think for yourself' strategy follows a central idea of third-wave feminism in general and Riot Grrrl in particular—that feminism can and should be reworked to be relevant to individuals' specific realities.

Volunteers also emphasize the importance of taking generational change into account. While older punks can take on the role of expert to legitimate their continued connection to a youthful local scene—teaching younger generations about past history framed as more authentic than current versions of punk (Bennett 2006; Davis 2006)—women at the rock camps share their experiences but also strive to recognize how girls' lives may be different from their own pasts. They highlight the importance of making the relationship between generations a collaborative model rather than a student–teacher relationship. Donna underlines the importance of the camp program being a 'kid-led activity', specifically because she feels that 'what I experienced as a young girl . . . is different from how they're living now'. She recognizes that although girls still face difficulties during adolescence, the specific problems and cultural context are different from her own experiences as a young girl, and consequently she does not want to impose an adult-informed view of the world on the girls as they do things such as write their own songs. As Becca, an instrument instructor at GRSW, explains, 'We have a lot to teach the girls, but it's important that they know that we're not there to tell them what to do or to be authority figures.' Even when they are doing the formal teaching of how to use instruments, many volunteers do not want the girls to feel confined by traditional instruction and technical know-how. They teach basic skills while still allowing girls to get creative with their instruments. Tobi, cofounder of GR!M, explains the importance of this leeway:

> I don't hold my drum sticks the 'correct' [makes air quotes with her fingers] way because, otherwise, it makes it difficult for me to play how I want to. I don't want to tell the girls that they are holding their [drum] sticks wrong because I don't want to discourage them from trying things out for themselves, in a way that might be more comfortable to them, but I can make suggestions for how to help them.

Highlighting their commitment to the Riot Grrrl philosophy that 'you don't have to play to play' (Molly, GRSW), Tobi and other volunteers instruct rock campers on the basics of their instruments while still allowing room for creativity in how and what the girls might play.

Organizers repeatedly emphasize giving campers strategies and tools for developing their own opinions, especially about what might be considered feminist. During a workshop at GRSW, Allison, a board member and counsellor, shows the campers the music video for the song 'Bad Romance' by musician Lady Gaga and asks them to discuss whether or not they think it is feminist. To facilitate the discussion, she draws the outline of a suitcase on a whiteboard and fills it with questions such as, 'Who's in the video? Who isn't? Who's speaking? Who's silent? Who's looking? Who's being looked at?' She calls this a 'feminist critical thinking toolkit', and tells the girls how this toolkit includes 'questions that I've found helpful to make sense of different situations I've been in'. She quickly adds, 'We [camp volunteers] want to give you some tools to analyze the media for yourself, and with these tools, you might develop different ideas and arguments.' Allison tells the campers that there are many different ways to be a feminist, and 'you don't have to agree with us [volunteers], and we don't want to tell you what feminism is . . . We're here to give you tools so that you can make your own decisions about it.' In this way, volunteers refuse to be *the* authority on feminism—punk rock and otherwise—and instead prefer to act as mentors who help to guide, but not determine, the girls' thoughts and actions. This way of teaching recognizes girls' agency to think critically and highlights the importance of dialogue between generations.

While many volunteers feel that the presence of positive adult role models plays a central role in helping the rock campers, they also acknowledge the mutual respect and learning that develops between the girls and women during the week. Volunteers, for instance, eagerly share the stage with these budding musicians. During a lunchtime band performance at GRSW, the lead singer of the visiting band asked the rock campers, 'Does anyone here like to scream?' and invited them to come on stage to scream into the microphones as the band played its final song. At GR!M, a band composed of volunteers played during a lunchtime performance and decided to cover a song written by a rock camper band the previous year. Many of the campers in that band had returned for that year's camp session, and they danced in front of the stage excitedly as their mentors played a song they had written. Therefore, although there are age divisions between campers and adult volunteers, these divisions do not necessarily lead to the traditional age-based assignment of teaching and learning roles.

New Models of Being a 'Grown Up'

Rock camps provide locations for an explicit transfer of knowledge between women and girls. Yet organizers also feel that girls may pick up implicit messages about the trajectories of adulthood—trajectories that can include a continuation of subcultural participation well past the teenage years. Organizers and campers frequently note how the people who volunteer at rock camp, as Tegan describes it, 'don't look like the mainstream'. Many volunteers do not look conventionally feminine[5]—or, as Donna describes it, 'We don't look like models in *Glamour* magazine.' It is quite common to see volunteers with facial piercings (e.g. lip, septum, eyebrow), gauged earlobes, short (and sometimes dyed) hair and visible tattoos—usually on their upper arms. Clothing choices vary, but a common outfit for volunteers includes lace-up sneakers, jeans that have been cut off at the knees and T-shirts with a local band name silkscreened on the front. The diverse looks of these women do not always match the cultural norm that adult women should appear feminine. Celebrating these appearance styles also challenges the assumption that participating in music scenes and fandom is not 'right and proper' for women past their teens and early twenties (Vroomen 2004: 242).

Volunteers see this diversity in gender presentation and visible continuation of subcultural participation into adulthood as a positive and important model for the girl campers. Donna explains how volunteers 'look different than the women they [girls] are exposed to on a daily basis—their mom, their teachers, and their parents' friends. I think it's really great for all these campers to get to see another side of things, that there is this whole world out there of women doing things differently.' Sometimes girls are unprepared for this display of gender nonconformity. Beth, a counsellor at GR!M, recalls getting complaints from a camper about the large tattoo on the side of her band coach's neck. 'She was put off by what she saw as the severity of her look,' explains Beth. Although some girls might initially be put off by such subversive appearances, many girls begin to embrace and even admire the nontraditional looks of volunteers.

Many organizers also feel that by interacting with the girl campers, they are complicating the idea of what it means to grow up and 'get a real job'. Tracy describes her hope that her presence at camp will help rock campers gain a new perspective on adulthood, explaining how campers 'will see that a lot of these volunteers are their parents' age, but that they are living differently while still being responsible and cool'. Showing how some campers make this connection, Allison, a board member at GRSW, recalls a conversation between herself and rock camper Erin:

> She was just talking about different things she liked, and at one point, she said, 'One of the things I really like is childish adults,' and she sort of looked in my direction. So I jokingly said, 'Hey, watch out! Are you talking about me?' and she goes, 'No, no. I mean, you're a grown-up, but you still seem like you have fun and you like to laugh' and stuff like that. I felt like it was just this moment where she was sort of saying, 'Oh, I see that you're having this style of being a grownup that I like,' and for her the term for that was 'childish grown-up'.

The women who volunteer at rock camp do not necessarily look and act like the adults who are part of these girls' everyday worlds—something on which many camp volunteers pride themselves, and see as an essential part of their roles as youth educators—specifically, that they are presenting rock campers with positive examples of different ways of being an adult. Kathleen distinguishes girls' rock camps from what she describes as more 'traditional' girls' organizations, which she argues 'give girls a narrow image of "success" as having an established career as a doctor or a lawyer'—the dominant version of adulthood that confronts middle-class youth. This version of adulthood does not involve continued subcultural participation. Allison elaborates on her belief:

> There's this narrative that's like, 'Yes, you can have your subculture when you're young, but when you grow up, you need to get a job and get married and direct all your energies towards your home and career.' But at camp, they [campers] can see all these people who are having a different way of living.

Organizers also complicate the assumption that marriage and children are the inevitable and only pathway for women. When Kathleen volunteered at a co-ed music program for youth, one of her guitar students asked her, 'Are you one of those women who travel?' 'Basically,' says Kathleen, 'she wanted to know if I was married or not.' The student was surprised to find out that Kathleen had never been married, because it went against the typical model of adult womanhood with which she had been presented. Although several volunteers are married and/or have children, most volunteers are not married and do not have children; additionally, over half of the volunteers interviewed identified as queer. Queer-identified volunteers, such as Kaia, a board member at GR!M, feel that it is important for them to 'be in this space and be queer . . . if my presence as a queer person helps even one camper, then I know it [volunteering] is worth it.' In fact, volunteers shared several stories about girls 'coming out' during camp week and how they are able to offer their own experiences and to support them—something that many queer and trans youth are lacking in other spaces that assume and privilege heterosexuality and cisgenderism.

Similar to many queer subcultures (Halberstam 2005; see also Chapter 2 in this volume), the adult volunteers and organizers are 'queering' the typical (and heteronormative) model of adulthood: get a job, get married, have children. Further, they are complicating assumptions that deeply held and visible connections to the music and ideologies of subcultures must fade in adulthood. These women not only continue to make rock music, but also 'look the part'. Volunteers show girls that there are many life pathways—that motherhood and playing in a punk band, for example, are not mutually exclusive. They also make visible queer and trans lives for girls who may never have had access to such possibilities. As Allison explains, it is important 'to give them [girls] a sense that there are many ways and many different timelines by which you can live your life, not just the dominant one that you hear about all the time'. Emphasizing the importance of young girls seeing this kind of subcultural continuation from adults, she notes, '[It] lets the campers see this subculture and that all these women musicians are still involved in it and that they didn't have to give it up [in adulthood].' Showing their perspective that punk rock feminism can be an identity rather than a youthful phase, volunteers fully expect that these camps will have a similar kind of continuity. Lori, an instrument instructor at GR!M, says her favourite part of rock camp is 'seeing girls transition from being campers to interns to instructors. It's awesome. You know they get it. You know that in thirty years, this is still going to be happening because it's so powerful and awesome that people are going to be doing it forever.'

Institutionalizing Riot Grrrl Politics

What is unique about girls' rock camps compared with many other subcultural music scenes is that they are institutionalized as not-for-profits. This institutional location formalizes the intergenerational transfer of knowledge that can occur informally in local music scenes. Yet creating an organization based on a punk rock feminist ethos can constrain how volunteers present their alternative identities and politics. Rock camps must present a public face that not only appeals to funders, but also to parents—who must be willing to leave their children with tattooed and pierced women wielding guitars. Such appearances can be overlooked in the name of 'rock'—parents expect rock camp instructors to look like rockers. However, punk rock feminist ideologies can be harder to sell to parents and grantors. Some organizers feel they have to temper their personal version of Riot Grrrl—a subculture premised on exposing misogyny and hypocrisy in explicit, controversial ways. As Kathleen laments, 'We can't be a 100 per cent Riot Grrrl camp.' Unlike Riot Grrrl, which established a subcultural space separate from adults, rock camps must take into account parental perceptions. As Hannah explains, 'We can't be in your face with our politics like Riot Grrrl was because we need our camp to be approachable to different types of parents who might not want their kids to hear about gender or how gay people are OK.' In order to be seen as approachable, organizers frame the purpose of the camp as promoting 'self-reliance and strengthened self-esteem through music'. This framing reflects larger societal concerns about girls that have surfaced in recent years—that girls' self-esteem is plummeting and that corrective action must be taken to give girls their 'voice' back (Gilligan and Brown 1992; Pipher 1994). In doing this, organizers take advantage of the mainstream popularity and acceptability of 'girl power' rhetoric, which now is largely stripped of its radical feminist roots (Schilt 2003b). Myra, a counsellor at GRSW, jokes that the camp is not dropping feminist politics, but rather modifying them so that rock camp is 'the slightly sweeter version' of Riot Grrrl.

Some volunteers, however, are concerned that taking up such an apolitical and individualized version of Riot Grrrl politics will blunt the political impact of the organization. While these women are not trying to pass on the 'real' meaning of Riot Grrrl, they also fear having to present a version of punk feminist that relies on empty slogans rather than strong ideology. Organizers navigate a tightrope as they hope to make the camp more appealing through discourses of 'girl power' while avoiding falling into a commercialized and individualized version of it—that is, they want to reappropriate the political potential of the phrase that has been so meaningful in their own lives. One example of how organizers struggle with using the language of 'girl power' (and its equivalent, 'girls rock') is reflected in a debate over the names of these organizations. At GRSW, for instance, there was a debate over whether they should use the phrase 'girls rock' in the name of the organization. Disagreements emerged, as Hanna describes it, 'mainly because the term has been commodified. I can go to Michaels [an arts and crafts store] and buy something that says 'Girls Rock!' on it, and it's not very revolutionary.' Another board member at GRSW, Christina, points out that although the phrase is commodified, she does not think it is entirely a bad thing: 'In a way, it's good that something like "girls rock" is commodified and out there more because it helps us attract more people.' For her, this slogan has the potential to attract a wider audience of people who are familiar with and can get behind the message of empowerment embedded in this concept. The mainstreaming of the Riot Grrrl ethos in phrases such as 'girl power' consequently becomes a site of contestation among volunteers as they balance the presentation of their Riot Grrrl identities and politics with the needs of the organization: some encourage drawing on the mainstreamed version of the Riot Grrrl ethos in order to attract a wider audience, while others want to avoid depoliticizing their punk rock feminist politics.

Just as 'girl power' and other Riot Grrrl principles were co-opted and watered down once they entered mainstream culture (Schilt 2003b), several organizers expressed concerns that the idea of a girls' rock camp might be co-opted and depoliticized as well. Tracy said of rock camp:

> It does have the potential to become very commercialized and lose the Riot Grrrl-feel, that going against the establishment feel . . . so it could go from being political to being an *American Idol* camp. That would be what we want to avoid.

Several volunteers pointed out how these camps need to work to bring 'girl power' back to its radical feminist, punk rock roots. Allison suggested that they need to actively rework the meanings of this phrase:

> This generic 'girls rock' message feels separate from an analysis of gender and the real barriers that they [the campers] are going to encounter as girls and women. I don't know how helpful it is just to tell them, 'girls rock' because that's really just like saying that girls can do anything without telling them, 'Here's what you might expect from people who think that's not true.'

For Allison, without reinfusing phrases such as 'girls rock' with a punk rock feminist politics of resistance, the camp is just recreating a slogan with an individual and apolitical meaning. For many volunteers, the muting of the punk rock feminist message could end up signalling the demise of the Riot Grrrl-inspired organization that helped to create it, as well as their strategy for continued involvement in this scene.

Conclusion

Lamenting the dearth of research on girls' subcultural participation in the 1980s—as well as the gender socialization that limited girls' ability to deviate from traditional feminine norms—Michael Brake speculated on how transformations would emerge: 'Changes in girls' attitudes will come from the influence of an older age group, and through the medium of feminism' (1985: 183). In many ways, Brake's hypothesis has been realized in girls' rock camps. Within these settings, adult women interact with younger generations of girls, teaching them not just how to use instruments, but also why taking up space in male-dominated locations is important and valid. Rock camps also provide an institutional location for intergenerational mixing, allowing women and girls to participate together in a subcultural space that is not divided into age-based roles of teacher and pupil.

Comparing rock camps to previous work on ageing subcultures suggests some gender differences in strategies for continued participation in punk scenes. For some ageing male punks, there is an effort to distance themselves from activities that seem age-inappropriate, such as being at the front of a stage moshing in front of a band

(Bennett 2006; Davis 2006). These men, in contrast, negotiate new roles that mimic the relationship between elders and youth, such as educating young punks about the past. This role also cements their legitimacy as older people in a subculture by making them, by virtue of age, the arbiters of punk authenticity. Women in the rock camps also engage in educating new generations. However, drawing on their feminist ideologies, they seek to create an intergenerational dialogue on the experiences of women and girls. They provide girls with critical thinking tools about Riot Grrrl, feminism and misogyny, as well as technical skills to increase the visibility of women musicians. Their hope is that this strategy will allow the girl campers to build their own ways of seeing and doing. Women volunteers also engage in 'youth' activities, yelling and screaming as fans at the girls' shows, complicating the idea of these forms of participation as the domain of the young. Finally, these women provide alternative models of adult womanhood. They show not just that women can still look punk and be in bands, but also that marriage and/or a job as an executive are not the inevitable and only careers available to women. These models—largely missing from popular culture—are a necessity for effecting social change in the lives of girls and women.

Examining the threads of connection between Riot Grrrl and girls' rock camps illustrates the need for a deeper investigation of women and ageing subcultures. Riot Grrrl offered an alternative idea of what it meant to be a woman. The interviews with the volunteers and organizers of two girls' rock camps highlight the impact that such an alternative view had on their lives as they escaped the evaporation of enthusiasm predicted by the press writing about the Riot Grrrl phenomenon. These findings support Angela McRobbie's (1991) claim that subcultural participation has 'the capacity to change the direction of young people's lives' (30). These women retain a dedicated commitment to using DIY cultural production to express feminist ideals for themselves and for a new generation of girls. In other words, they did not 'grow out' of punk rock feminism, but rather found a new avenue for expressing their ideologies. Keeping the Riot Grrrl ethos alive for a new generation is not without its own conflicts and contestations; however, this model of subcultural adults creating organizations for transmitting their knowledge to young girls—many of whom are not affiliated with punk rock cultures—presents a new model for subcultural participation beyond the teenage years.

–12–

Parenthood and the Transfer of Capital in the Northern Soul Scene

Nicola Smith

Introduction

The British Northern Soul scene in many ways epitomizes youth culture, with its underground dance clubs, recreational drug use, all-night dancing and in-crowd exclusivity. Alongside such typical youth cultural practices, scene members have forged identities to demonstrate their competitive connoisseurship, their subterranean status, and their passion for and dedication to the scene. Yet, five decades after the cementing of such practices and identities, the same scene members continue to demonstrate this impassioned performance of Northern Soul fandom—but as mature scene participants. In the words of Curtis Mayfield, the ageing soulies *keep on keeping on* in middle-age, in old age—and in parenthood. This is the story of the soul parents. This chapter discusses matters prevalent to all music scenes with longevity and those with an ageing fan-base via discussion of adult influence (rather than youth resistance) in popular music and the familial (rather than peer) inheritance of subcultural capital.

The Northern Soul scene offers an unusual opportunity to explore distinctions and similarities in youth culture practice across generations. Originating in the late 1960s, the scene has maintained founder members and has (largely) been accepting of newcomers throughout its five-decade history. Some of these young newcomers are the children of existing and previous participants—these I term 'soul children'. I have argued elsewhere that there is a variety of participant types on the Northern Soul scene today, and that there are potentially different stages to scene participation (Smith 2009). The current scene consists of established participants (those who have been on the scene continuously and for several decades in many cases), returning participants and newcomers, mature and young.

Within the example of Northern Soul, we are witness to the implication of inheriting (sub)cultural capital, not via social class (Bourdieu 1984) or peers (Thornton 1995) but via micro-familial inheritance. This chapter presents the thesis that cultural participation is comparable to a family heirloom. The soul child does not merely inherit a record collection but a cultural identity; he or she inherits passion for a scene, comprehension of scene participation and the ability to perform connoisseurship via parental reminiscences, parental example and passed-down commodities. This inheritance allows easy access to scene membership. Put simply, the introduction to popular music culture that a parent[1] may provide for his or her child is not necessarily confined to the household environment. It has extended to the record stalls, DJ booths and dance floors of an active music scene. Such *inherited familiarity* with a scene and its music raises questions about parental motivation for sharing cultural spaces with offspring. Likewise, the appeal of the scene for the soul child requires consideration.

The data presented within this chapter are drawn from ethnographic research, including interviews with fans and participant observation of the contemporary Northern Soul scene since 2005 across the United Kingdom. Brief questionnaires were also completed by research participants to gauge background information (length of time on the scene, number of children and so on). The people interviewed were self-proclaimed Northern Soul fans, located either via their scene involvement (event organization, fanzine writing, online presence, DJing) or via face-to-face encounters on the scene. In total, 91 participants directly contributed to this research.[2] I will draw upon their stories of fandom throughout; however, for the purpose of this chapter, emphasis will be placed on four specific case studies. For ease of definition and for purposes of anonymity, I have termed the four case studies: Mike, the father; Frank and Audrey, the couple; Daniel the son; and Jack and Jill, the father and daughter.[3]

The Northern Soul Story

The term 'Northern Soul' was coined by journalist Dave Godin,[4] who realized that a preference for early and mid-1960s rare, obscure American soul (sounding like Motown but usually on small, independent record labels or lesser-known subsidiaries) existed amongst customers from the north of England who visited his London record shop, Soul City, which he opened in 1968. The term was used to illuminate the distinction between the soul record-buying customers from the north and those from London and the south, thus the label Northern Soul signified a distinction between record-buying habits of UK soul fans rather than alluding to music from North America. As Godin explains:

> I started to notice that Northern football fans who were in London to follow their team were coming into the store to buy records—but they weren't at all interested in the latest developments in the black American chart . . . I devised the name as a shorthand sales term . . . It was just to say, "If you've got customers from the north, don't waste time playing them records currently in the US black chart, just play them what they like—'Northern Soul'." (quoted in Hunt 2002)

The scene experienced a peak in popularity during the mid-1970s. This peak is often associated with the Wigan Casino.[5] Northern Soul experienced a relative decline in popularity throughout the 1980s as several pivotal venues closed, namely the Highland Room at the Blackpool Mecca in 1979 and the Wigan Casino in 1981. Some participants left the scene in response to what they considered the (negative) mainstream attention Northern Soul had experienced in the 1970s. Furthermore, by the early 1980s many participants had started families. The added social and monetary responsibilities of having a family dictated that many participants left the scene to concentrate on work and family life. This brought about a lull in the scene, with fewer major venues, a decrease in regular events and a decline in the number of regular attendees at the events still being held. As a result, Northern Soul arguably reverted back to an underground standing in the mid-1980s. Once again ignored by the mass media and by the public majority, the scene was kept alive by word of mouth and the continued passion for the music by a relatively small number of existing fans. Although event frequency severely reduced during the 1980s, venues such as Stafford's Top of the World, London's 100 Club and Morecombe Pier ensured that the scene remained active. The mid-1990s saw a scene rebirth that continues to the present day. Many fans chose the mid-1990s to return to the scene because, for many, their children had grown up and left the family home. More free time and greater disposable incomes enabled fans to return to the scene—which was still in existence. Today the scene has overcome a decline in popularity and, in addition to its wave of returning fans, has attracted new blood and fresh legs as the infectious sound of 1960s soul has swept a younger generation on to the dance floor.

Becoming Soul Parents

For several decades, ageing soulies have been performing their fan identities. This performance of Northern Soul fandom is a display of a participant's understanding and knowledge of the scene. Often performed competitively, fandom is shown within displays of knowledge of the music, dance moves, people, venues, record labels and so on. Ageing on the scene is something of a double-edged sword. On the one hand, longevity leads to experience and a honed performance. Peer adulation correlates with time spent on the scene. On the other hand, with age comes reduced physical mobility and, for some, increased social responsibilities. Such factors imply that aged participants within a youth culture will cease participation, but to varying degrees these issues have been overcome by a dedicated body of ageing soulies.

The impact of physical ageing has been contained by the promotion of experience beyond physicality within performances of fandom. For example, one way of containing the negative impact of ageing is via the positioning of time spent on the scene as a sign of fandom. Dancing in a knowledgeable, practised fashion, alongside people who have known you for many years, has become significant—even if that dancer cannot perform dance moves with the stamina they once exhibited in their youth. Second, the fact that the continuing scene is frequented by adults—rather than youth—counters mainstream assumptions of age-appropriate behaviour within music cultures. Finally, although the core of social responsibility tends to be situated within domesticity and family life, the ageing soulie negotiates the apparent boundaries drawn between leisure activities and home life by incorporating parenthood into the practices of fandom. As I have argued elsewhere:

> Hedonistic exploration and the use of leisure to inform self and to achieve imagined escape do not necessarily have to cease as a consequence of maturity. The explorative necessity may become exhausted in adulthood, yet the identity acquired via this initial youthful exploration informs and continues to inform selfhood and personhood if allowed. The . . . behaviours practised within the scene are, in the main, the same today as always . . . Youth actions originally considered as distinct from adult society [are] being performed by adults. (Smith 2009: 435)

Interestingly, this performance of fandom is often occurring alongside parenthood. Of the scene participants I surveyed, 56 per cent had children, and 63 per cent of them said that they listened to Northern Soul *with their children*. The format of this listening and the space/place in which this occurs varied widely. Some listened to Northern Soul in the car or at home, where 'it just so happens the children are' (male, 54). Others listened to Northern Soul with their children in a more purposeful manner, with one participant suggesting that he used to nurse his daughters as babies to Northern Soul records (male, 47). Shared listening also occurs at Northern Soul events. Several people take their children to events, with one fan stating that 'it's nice, all the family going together rather than mum and dad leaving the kids at home' (female, 59). Several fans suggested that their influence can be recognized within their children's musical preferences, with one fan proudly stating that his daughter dances at Northern Soul venues. He likes to go along because it 'reminds him of the good old days' (male, 45), suggesting that his daughter's involvement in the scene is an encouragement for his own continued fandom. One soul parent stated that his daughter knew how to dance in the Northern Soul style but wouldn't do this 'out of the house, it's just something she does at home, mucking about' (male, 42), and another commented that his 19-year-old daughter had 'liked soul for the last couple of years' but admitted that 'she won't go out to listen to Northern' (male, 46).

From the above comments, it is clear that although soul parents provide their children with an awareness of Northern Soul, not all of these children automatically become Northern Soul fans. One DJ with whom I spoke suggested that it was not the Internet that was keeping the scene known and alive, but 'mum teaching her daughter the moves to "Do I Love You"' (male, late fifties).[6] Clearly there is a variety of modes of soul child musical consumption, and varied responses from soul parents to this consumption. The following case studies highlight the array of soul parent positions and practices.

CASE STUDIES

Mike, the Father

Mike is aged 47. He has two daughters (aged 13 and 15) who have 'grown up around Northern Soul'. Despite playing songs in the home and telling his daughters about the scene and record labels as he has known them, he is adamant that he does not want his children to become soul children (i.e. to become on-scene participants). He has never encouraged his daughters to attend soul events, nor does he expect them to do so when they are old enough. He struggles to see the appeal of children attending events with their parents: 'The girls know of Northern and . . . I like that . . . they've been singing along [to Northern Soul songs] for years, all their life', but he suggests that his daughters know that it is their father's scene. He is still active on the scene and does not imagine having to share it with his offspring in any direct way. In this example, there is a sharing of the music but within the domestic (rather than leisure) environment.[7] Mike has promoted scene familiarity but only within the context of the children's recognition of their father's fandom. There is no aim to introduce his offspring to his cultural practice outside of the home. When asked if he would feel the same if he had sons not daughters, his response was the same. I have written elsewhere of the conflict of age evident on the Northern Soul scene (Smith 2009), in terms of the struggle that some ageing existing participants are experiencing in response to younger scene members. For some, soul children—with familial scene connections—are a less threatening group of young newcomers than those who have no previous connection to the scene; however, for others, any younger participants equate to a threat to the ageing soulies' scene ownership. Incorporated within this is the issue of the internal conflict of the ageing self: ageing soulies are faced with their own reduced physicality occurring alongside visual displays of stamina and energy from dancing young newcomers. It could be argued that Mike's clearly drawn line between his daughters' domestic consumption of Northern Soul and the out-of-bounds external scene reflects negative feeling towards a youth presence exhibited by some of the ageing Northern Soul community. Mike demonstrates a negativity towards any youth presence that many (although, it is vital to state, not all) older soulies exhibit, as well as a personal desire to protect his own social space/identity vis-à-vis his own children.

Frank and Audrey, the Couple

A married couple, both now in their early sixties, Frank and Audrey have been into rare soul and black American music since their teenage years. As with many Northern Soul fans, Frank and Audrey have continued to participate in the scene throughout adulthood, and consequently the music has become an aspect of their cultural identity. Northern Soul holds nostalgic appeal for the couple, but also informs their social world, their family life and their personal relationship. It is clear that Northern Soul is a shared passion between them, and they exhibit this passion in both a domestic and a leisure environment. Several of their friendship groups are built around Northern Soul fandom; the couple socialize with those on the scene and take soul-related holidays with fellow soulies. They have two daughters: 'One loves Northern, . . . the other doesn't but her boyfriend . . . love[s] the scene' (Audrey). The parents have always played Northern in the house, but 'more for ourselves than for [the daughters]'. They know their daughters are aware of Northern—with one daughter who 'takes my CDs to copy' (Frank)—but they suggest that they did not set out to 'teach' Northern Soul; their children 'just kind of picked it up' (Audrey). They are neither pleased nor offended by their daughters' interest (or lack of interest) in Northern Soul, but did state that the daughter who liked Northern 'would not come [to a Northern Soul event] with us . . . she likes the music but she does her own thing' (Audrey). Both daughters can dance in a Northern Soul style—'they've done it since they were kids' (Frank), and the daughter who is into Northern does dance in this style at clubs but not with her parents. If their daughters asked to come along to a Northern night, the couple would 'be surprised but . . . wouldn't say no' (Audrey). As yet, neither daughter has asked.

Daniel, the Son

Daniel is aged 20. He has been familiar with Northern Soul 'for as long as [he] can remember' and has been participating on the scene since he was approximately 12 years old. He would attend events with his parents when he lived in Doncaster, but he has recently started university in Scotland so goes to soul nights with a small group of friends there. He also attends modern soul nights and has a wider passion for obscure soul and rare vinyl. His father was a collector of Northern Soul records, but he has not been active on the scene for roughly three years now. The father hears of the scene from Daniel. Daniel thinks his father 'likes to keep in-the-know [about the scene]. He likes to know which records are hot . . . I have his record collection but he still loves the music.' When asked why his father gave away his collection, Daniel suggested, 'I guess he didn't need it anymore.' Daniel was given (and has since developed) his father's record collection. Daniel now trades and buys records online and has 'tried a little bit' of DJing but 'nothing too flash . . . just for my mates really . . . and one spot at uni'. Daniel DJs using his father's record collection. He participated on the scene while his father was still an active member. Daniel does not dance, and neither did his dad. In this example, cultural inheritance is demonstrated by heirlooms (records) as well as scene knowledge.

Jack and Jill, the Father and Daughter

Jill was 18 when I met Jack and Jill for the first time; Jack was in his mid-forties. Jack doesn't live in the family home anymore (he is divorced from Jill's mother). Home is therefore no longer the site for cultural inheritance (from Dad) that it once was. Jill is familiar with Northern Soul because of 'Dad playing Northern around the house' and thus she has known the music since childhood. Now Jack spends time with his daughter at soul nights; he described it as 'a thing we have' and explained that Jill 'didn't really' attend Northern Soul events before the divorce (she was 16 when her parents separated): 'I took her to a few weekenders when she was younger, her mum was kind of into the music then so it made sense to bring Jill with us.' Jack attends events without his daughter but knows that 'the odd soul night' is a place where they can 'do something together'. Jill attends some soul nights without her dad but infrequently: 'I know the music [and] it's kinda cool that people are impressed I know all of this old music and [that] I can dance a little,' she suggests, adding, 'The retro thing is quite big with some of my mates.'

Passing Down Knowledge: The Subcultural Capital Exchange

Borrowing from Pierre Bourdieu's (1984) notion of cultural capital, Sarah Thornton notes the existence of what she terms subcultural capital (1996) within club cultures. Thornton identifies 'subcultural ideologies', which are 'a means by which youth imagine their own and other social groups, assert their distinctive character and affirm that they are not anonymous members of an undifferentiated mass' (10). Thornton subscribes to Bourdieu's notion that cultural capital equates to the accumulation of knowledge; Bourdieu envisaged this accumulation occurring in the home and via formal education, while Thornton allows for 'subspecies of capital operating within other less privileged domains . . . the terrain of youth culture' (11). Thornton's subcultural capital is useful when exploring the Northern Soul scene in the sense that participants do accumulate a form of scene-specific knowledge. This is a performance of 'hipness' for Thornton, with participants recognizing how to act, how to dress, what to listen to and what to say in a given social context. However, Thornton is positioning subcultural capital within a dichotomy of mainstream versus subculture, with adulthood a component of the mainstream, and thus subcultural capital is considered the domain of youth *only*:

> Going out dancing crosses boundaries of class, race, ethnicity, gender and sexuality, but *not* differences of age . . . A loss of interest in clubbing coincides with moving out of the parental home, which has repercussions for young people's desire to get out of the house and escape the family . . . Clubs allow their patrons to indulge in the 'adult' activities of flirtation, sex, drink and drugs, and explore cultural forms (like music and clothes) which confer autonomous and distinct identities. (Thornton 1996: 15–16, original emphasis)

Unlike the club cultures Thornton witnessed, the current Northern Soul scene *does* cross boundaries of age.

Many soul parents have honed their subcultural capital for decades. As they age, there is the fear of this subcultural capital 'going to waste' (Daniel). The act of passing down scene-relevant knowledge to offspring so that the child can translate this knowledge into subcultural capital is appealing to the soul parent for two reasons: first, their involvement in the scene continues (vicariously through the child), irrespective of a reduction in physicality; and second, subsequently the scene gains fresh interest from a young(er) audience and thus continues.

In the example of Northern Soul, subcultural capital can be achieved *in the family home*. As such, parental influence is positioned as a source of youth's cultural knowledge rather than parental musical taste presenting a barrier to intergenerational interaction. Although I would resist the urge to draw the conclusion that in Northern Soul we have an example of a music culture situated beyond generational boundaries, it is worthy of note that generational distinctions in musical taste can become *shared* musical tastes, and such sharing can be beneficial within the family unit, as we see in the example of Jack and Jill, who are dealing with divorce.

The capital that soul children achieve is somewhat akin to Bourdieu's original concept of cultural capital, achieved in the home (but not via formal education). However, in the example of Northern Soul, the potential soul child learns not of sociocultural convention as Bourdieu suggests, but of subcultural practice. This is therefore not a dichotomy between the adult mainstream and the youth subculture, but is instead an age-fluid *inherited subcultural capital* that is learnt in the family home and later taken to a more traditional subcultural space (a nightclub, for example). Parental subcultural capital is converted into inherited subcultural capital. Once the child of a soulie takes music from the family home and translates this into scene participation, he or she has become a soul child. The child learns of Northern Soul but also learns of his or her parent as a soulie via this subcultural capital exchange. For the parent, the child's recognition of the scene and of his or her parent's place within it is arguably an extension of the soulie peacock display. It is a method of gaining attention as a fan, but in this example it is offspring recognition rather than peer recognition that is achieved.

The concept of socialization is of relevance to the story of the soul children.[8] The soul child is socialized into comprehending specific cultural practice from an early age. The children cannot *un-know* such cultural practice and objects of fandom as introduced to them by the socializing agent that is the family, but they do have the option to accept or reject continued engagement with Northern Soul when they grow older. The option to choose is shown clearly in the example of the two daughters in the couple case study: one daughter opts to translate her childhood introduction to Northern Soul into soul child participation, whereas the other daughter opts to reject the scene despite growing up in the same musical environment as her sister. This indicates the agency of the child and reminds us that the soul parent does not (always/often) intend to initiate their child into Northern Soul fandom. A soul parent performs his or her Northern Soul identity at home via domestically situated consumption of the music. This is a performance, whether conscious or not, and spectators are influenced by this performance. In the domestic setting, the audience can be a child. Soul parents embrace the chance to display their knowledge to influence others and simultaneously aid the continuation of their scene. Interestingly, this sharing of knowledge and influence is not primarily scene-based: it occurs in the home.

From Home to Scene

In many music cultures, youth typically are considered to travel to spaces beyond the home to engage in music, thus performing beyond the watchful eye of their parent(s).[9] However, in relation to Northern Soul, the home is the domain of the ageing soulies. No longer are soulies listening to Northern Soul music in private spaces within their parents' home, but this music is now an open soundtrack within their own home. Music that the soulies once played in their bedrooms as teens is now being played in the family car, the kitchen and the living room.[10]

The metaphor of the home is often used to explain the concept of fandom. For example, Will Brooker (2007) writes of 'a sort of homecoming' to symbolize fan pilgrimage and Cornel Sandvoss (2005) suggests that 'fandom best compares to the emotional significance of the places we have grown to call "home", to the form of physical, emotional *and* ideological space that is best described as *Heimat*' (2005: 64–5, emphasis in original). Sandvoss suggests that *Heimat* is

> an imagined relationship between the self and the external world, in which part of the world is experienced as inherently related to and constituted through the self; it is one's place in the world, in which place and community become an extension of one's self, and the self a reflection of place and community. (2005: 65)

In the example of the soul parent, the imagined home of fandom is made more real by the presence of kin within the selected fan space. Moreover, soul parents have found another space (beyond the scene) in which to perform their fandom when they perform to family within the home. Thus soul parents can guarantee that, despite their personal ageing and the aged scene, they will always have a space in which to perform their identity as a soulie: the home.

Should a child opt to convert scene-relevant knowledge into Northern Soul fandom by translating inherited familiarity into scene participation, then the home becomes a space in which the soul parents can hear of the active scene via their child, thus living beyond their own active scene participation. As such, the soul parents' impact on the scene multiplies because their influence is felt in two performances of fandom: their own and that of their child. As Daniel tells us, when the day came that his father could no longer participate on the scene, then the family home became a space for scene attachment, fuelled by familial involvement and generationally shared objects of fandom (vinyl records).

Northern Soul consumption at home is significant for two reasons. First, it provides space for subcultural capital exchange from parent to child. Soul parents have the opportunity to enhance their attachment to the scene if their child becomes part of that scene. The inheritance of a social world not only provides a cultural outlet for the soul child, but enables the soul parent to forge additional levels of influence upon the active scene, meaning the scene continues due to additional newcomers (the soul children) and the fan identity of the soul parent simultaneously endures.[11] Second, the home provides an additional space for Northern Soul fan performance. The soul parent is provided with another (and most likely positively biased) audience, thus granting the ageing soulie a positive fan experience. Matt Hills (2002), discussing the influence of the family on fandom, notes that the family is one persistently privileged social grouping which can act to shape and organise the contingencies of the child's early (and continuing) object relationships, particularly where the cult text is encountered initially in the private sphere (on television, video, radio or in the novel) (163).

Whereas TV, radio or novels hold the essence of many forms of cult fandom in Hills's account, in Northern Soul it is scene participation that is the essence of fandom. Thus the home can provide knowledge of the music and the scene to some extent (in an anecdotal form), but the child does not become a soul child until he or she *participates* on the scene. Northern Soul in the home appears to provide an opportunity for child–parent bonding. When children opt to develop their attachment to Northern Soul beyond an initial opportunity to bond with their parents, we have a situation of familial shared cultural participation, something both Daniel and Jill have experienced.[12]

While various strands of cultural identity do not tend to intrude on one another, to be *simultaneously* a soulie and a soul parent—in other words, to experience on-scene participation with both peers and offspring—is potentially problematic, especially in a scene known for recreational drug use and all-night dancing. For this reason, many ageing soul parents suggested that they were pleased that their children enjoyed the music; however, while they saw value in this, they did not entertain the idea of sharing the scene with their offspring. The desire to keep scene participation and family consumption of Northern Soul separate is evident in Mike's account and in comments from the wider Northern Soul community.[13] The line drawn by Mike was to separate familiarity with the objects of fandom from participation within the active scene. He considered a participating child inappropriate and unwanted, and was keen to keep scene membership for himself. Such a separation has enabled Mike to maintain an undisrupted scene life while also providing a secondary space in which to perform his own fandom (the home). Thus, ageing members of this youth culture are negotiating their place in the scene, closing off certain parts of Northern Soul fandom to their children because they regard it as a domain somehow separate from the sphere of parenthood and the moral responsibility that parenthood brings.

The case studies show that there are varying degrees of parental encouragement for child scene participation. Of note is Mike's struggle with the possibility that his children will frequent the scene while, in contrast, Jack develops his relationship with his daughter via the scene. Mike wishes to keep Northern Soul with his generation while Jack embraces connections between the scene and family life. In the example of Daniel, we witness a connection between Daniel and his father created by a shared passion for Northern Soul, yet their story is not about on-scene participation but instead is centred upon the sharing of objects of fandom.

Objects of Fandom

Sandvoss (2005) suggests that objects of fandom become part of a fan's self, and vice versa. This narcissistic reading of fan self-reflexive engagement with his or her objects of fandom proves a useful paradigm by which to understand the pleasure that the soulie will experience when passing down objects of fandom to their offspring.

If the object of fandom is an extension of fan self then, when the fan wishes to detach from such objects (for whatever reason), to pass the objects to a part of the self (offspring) is both a continuation of that self-attachment to the object and, arguably, a less brutal and less final detachment from the objects of fandom. In fact, this could be considered not detachment at all, but simply a relocation of the objects of fandom. Sandvoss positions fans as

> fascinated by extensions of themselves, which they do not recognize as such. The notion of fandom as self-reflection describes the intensely emotional involvement between fans and their object of fandom . . . Such self-reflections are manifested on a number of levels: in fans' failure to recognize boundaries between themselves and their object of fandom. (2005: 121)

This notion of merging self with objects of fandom is emphasized in the example of the soul parent who brings his or her offspring into direct contact with the objects of fandom, and thus ensures a prolonged, secure attachment to fandom as we see in the case of Mike, who alerts his family to the existence of his objects of fandom. Mike shows that such objects are linked to notions of self to the extent that, at this stage, Mike cannot envisage sharing the active scene or giving away his objects of fandom. Mike passes down *knowledge* (subcultural capital), but does not wish to pass down the *artefacts* of his own fandom. While not evident in Mike's account, cultural inheritance *can* be both symbolic (subcultural capital) and physical (cultural heirlooms), and Daniel demonstrates this. When Daniel's father no longer wished to be part of the active scene, he passed down his cultural heirlooms to Daniel, resulting in Daniel's active engagement with the scene. Northern Soul is a record culture, so subcultural capital can be passed to offspring in a physical (in this case, vinyl) format as well as via learnt experience and knowledge.

While (sub)cultural capital can be passed down to the soul child, such cultural meaning is only realized *within* the scene. This presents a reason for parents to promote Northern Soul to their child(ren). Inherited vinyl is a way of continuing the scene (the records hold meaning on the scene, thus encouraging the owner to participate on the scene) while also aiding the stasis of the value of the records possessed.[14] Passing down vinyl—or indeed any object of fandom—to a receptive, knowledgeable younger audience increases the likelihood that those objects will remain significant (and valuable) to the active scene.

It is evident from the above examples that Northern Soul fandom has moved from the scene into the home as participants have matured and, as the lifecycle continues, fandom is finding another route from the home to the scene again: via the soul children. Both Daniel and Jill found the scene and new friendship circles via Northern Soul. In both cases, scene involvement started with the parents and has grown into independent forms of cultural consumption and fandom. Thus it is worth briefly considering the impact of cultural inheritance and soul child on-scene participation.

Cultural Inheritance: Soul Children on their Parents' Scene

Fandom could be considered an indicator of self which is open across generations. However, it could be argued that parents are symbolic of the mainstream, and thus present to youth a reminder of everyday dissatisfaction and sociocultural limitations (Lewis 1992). So why would children want to find fandom in the shadow of their parents? Perhaps we should consider the soul children as we consider football fans who inherit a cultural preference to which they consider there to be no alternative.

Leaning on Erving Goffman's (1971 [1959]) notion that identity is constructed via the performance of self to others in a social setting, to achieve a sense of self-as-fan, the soul child must step out of the home (and the shadow of their parents) and on to the scene as a participant in his or her own right. While in the minority, there are many cases of children who attempt to convert parental introductions into independent personal involvement. Several parents in the case studies encouraged this by taking their children to Northern Soul events. Observations of the current scene show that on-scene visibility of parent–child shared participation in Northern Soul occurs predominantly with younger children (pre- or early teens). These children attend all-dayers and weekenders, and they frequent the dance floor for a 'lesson in Northern Soul' during quieter daytime slots and before the established fans arrive.[15]

The older soul child, however, is akin to Thornton's clubbers, seeking subcultural capital and, just like the nonsoul child newcomers, engaging with the scene as a route to peer acceptance and cultural uniqueness, autonomy and identity (just as the soul parents once did). The soul child in this example wants to do something else with his or her knowledge of Northern Soul. Soul children have inherited scene knowledge and they want to convert this into subcultural capital to use to their own advantage, as Daniel demonstrated via the development of his inherited vinyl collection and his search for new soul venues. Northern Soul can also be a source of retro cool—something to which their peers do not have easy access, and something in which Jill found pleasure and peer-relevant kudos.

So what does this mean for the ageing soul parent? Just as Bourdieu positioned the family as a source of knowledge which is applicable to and recognized as valid by those external to the family unit, the subcultural capital that soul children inherit is relevant to more than just the family, but its relevance is only understood *within the scene*. Thus the soul children are bound to the scene if they wish to convert their familial-derived scene knowledge into subcultural capital and subsequent social recognition and cultural status.[16] Inevitably, soul parents will one day stop attending Northern Soul events. As yet, however, there is no pressure placed on the soul parent to 'step down' because the current scene accommodates mature adults far in excess of teenagers and young adults. Perhaps there is some comfort for the ageing soul parent in the knowledge that the soul child cannot abandon the Northern Soul blueprint and thus cannot entirely break free of the parental version of the scene.

Conclusions

Many soul parents raise soul children and provide them with the necessary cultural capital and resources to 'authentically' engage with the Northern Soul scene. In the case studies, we see the sharing of cultural capital both symbolically (Frank and Audrey share scene knowledge with their daughters; Jack provides friendship circles for Jill) and physically (Daniel receives his father's record collection). The family home is a pivotal environment for the forging of musical taste, and the influence of family upon musical preference is clearly evident within the example of Northern Soul. What I am arguing here is that the sharing of music within the family home is not merely of significance to the child who listens and then accepts or rejects this music, but this musical sharing has relevance for the parents who, as ageing music fans, find opportunity to rejuvenate, and to some extent extend, their musical fandom and the identity associated with that fandom.

There are two components to familial-shared fandom. The first is the cultural inheritance of objects of fandom and scene knowledge, given by soul parents to their children. In this instance, musical artefacts can be categorized as cultural heirlooms, and musical knowledge can be considered inherited subcultural capital. The second is the cultural continuation of a scene and scene-derived identity alongside (and perhaps irrespective of) age(ing). The first occurs primarily in the family home and the second primarily on the scene, but with the passing of time the domestic setting may replace the scene as the key site of fandom for the ageing soul parent. If so, cultural artefacts and soul child conversion of parental influence into scene participation will provide an extended connection between the soul parent and the active scene.

Music consumption shared across a generation is of benefit to the parents, who are able to extend their performance of fandom into the family home while also exhibiting the extent of their own fandom to their peers, thus achieving cultural continuation by bringing their child(ren) to events. Soul parents are satisfied that they have protected the scene and their involvement within it via the soul child.

Knowledge of the scene is passed from parent to child like a family heirloom. Soul parents have gifted their children with cultural capital that is not reliant upon the process of learning from peers (watching on the scene; practising at home) as all other participants—themselves included—had to do. Soul parents have schooled—consciously or not—their offspring in the rituals of the scene by encouraging their children to listen, dance and sing to Northern Soul music. The ageing participants have created a household filled with Northern Soul stories, memories and music, and have thus provided an introduction to the scene for their children. Soul child scene knowledge is not formed through blind introduction, dedicated practice and cultural archaeology as was the soulies' experience, but instead the child's access to an elite cultural world has been present potentially since birth. In light of the fact that the Northern Soul scene continues, the stories of parental participation in this cultural sphere are not merely nostalgic anecdotes to tell offspring but are instead methods of subcultural capital exchange. By becoming soul parents, the ageing Northern Soul fans share their cultural world, and in so doing aid the continuation of their scene and, importantly for them, strengthen and prolong their involvement in the ageing Northern Soul scene.

Notes

Chapter 1—'More than the Xs on My Hands': Older Straight Edgers and the Meaning of Style

1. All names are pseudonyms.
2. A brand of permanent marker.

Chapter 2—Performances of Post-Youth Sexual Identities in Queer Scenes

1. The CCCS approach to subcultures emphasized stylistic forms of youth cultural resistance such as teddy boys, mods, rockers, skinheads and punks that had been developing rapidly in Britain since the 1950s.
2. BDSM is the common abbreviation for bondage, discipline, sadism and masochism.
3. The pink dollar refers to the collective spending power of lesbian and gay citizens.
4. While all my respondents were cisgendered, it should be noted that the scene also accommodates many transgender and genderqueer people.
5. All respondents discussed occasional use of recreational drugs, and drug use is evident at scene events. The drugs most commonly used are ecstasy (methylene-dioxymethamphetamine), speed (amphetamine), acid (lysergic acid diethylamide) and marijuana.
6. There are some exceptions: in summer, for example, an event may begin as a picnic or pool party and then progress into a nighttime dance party (see Taylor 2010). Moreover, festivals such as annually occurring Pride Festivals which occur in major cities globally often consist of daytime events, in which scene members may participate.
7. Reasons for this vary, but may include availability of space, hiring cost, the desire for different scenery, licensing restrictions of venues, availability of appropriate sound reinforcement and staging, and whether or not a party is planning to accommodate sex or sex play on premises. Such a party would normally require special adult entertainment licensing permits.

Chapter 3—Ageing Rave Women's Post-Scene Narratives

1. The term 'rave' may appear dated to some readers. Nevertheless, this terminology is used to situate interviewees' narratives within a particular time and place (Toronto's electronic dance music scene circa 1994–2000).
2. Asking interviewees about their identifications as 'ravers' caused some confusion. Grrrl's experiences are the most extreme example of this slippage, for while she no longer considered herself an active raver (and thus fitted my research criteria), part of her reasoning was that 'raves'—in the strict sense of unlicensed 'underground' events—are exceedingly rare. Indeed, at the time of her interview, Grrrl was still involved in Toronto's electronic dance music scene (though less regularly and not as a raver per se, but as a DJ).
3. Asking interviewees whether they would attend another 'rave-type' event indicates my recognition that 'rave' might not be reflective of current electronic dance music scenes.
4. The term 'raver dance' is slightly awkward. Nonetheless, I have kept Serendipity's phrasing here as a way to simultaneously emphasize the extent of her embarrassment when telling me this story and the longing she described with recounting her efforts to fit into Toronto's rave scene despite her positioning as a relatively old(er) raver.
5. Following Hathaway (2001: 128), harm reduction is understood as an approach that 'urges safer drug-use practices as opposed to the elimination of all drug-use' and that represents, at its core, a deep commitment to human rights.
6. See http://www.tripproject.ca/trip for more information about this organization.

Chapter 4—'Each One Teach One': B-Boying and Ageing

1. B-boy: someone who breaks. Originally in hip-hop culture this meant a 'break boy', someone who 'gets down' (dancing) to the break of a record.
2. Battle: the competitive dimension of the dance. To battle someone is to treat them as an opponent whose skills are going to be compared to your own: whether this is judged or not, everyone knows that it is a competition.
3. See the interview with Afrika Bambaataa in the recent documentary *The Freshest Kids* (2002) for an example of how the 'elements' of hip-hop culture have expanded to include knowledge, peace, love, unity and having fun.
4. Personal correspondence, April 2011.
5. The high-rise apartment was located near the intersection of Jane and Finch Streets, an area in Toronto notorious for its crime and poverty.
6. Vietnam of the Rockforce Crew, who won best dancer at BOTY in 1999, and Lil' Tim of Airforce crew, who is a legend of the UK scene, are both top performers. Their emphasis on keeping learning fun and social for youth is significant

and matches what sports psychologists have argued lately about how integral friendly coaches and play are to the development of top performers.

7. At the 2011 'Movement' b-boy/b-girl event for Toronto's Manifesto Festival, Lance Johnson received the 'Most Original' Toronto Floor Award from his peers in the b-boy/b-girl scene.
8. ILL-abilities website: http://www.illabilitiescrew.com/about. Accessed 21 April 2011.
9. Mariano Abarca is widely known in the Canadian b-boy scene for his involvement with the Bag of Trix crew from the 1990s to the present. He was invited to speak at a panel discussion at Harbourfront Centre on 26 November 2011 as part of an expert panel.

Chapter 6—Rock Fans' Experiences of the Ageing Body: Becoming More 'Civilized'

1. The research also involved ethnographic fieldwork in the Northern Soul and rare soul and electronic dance music scenes with fans aged over 30.
2. In the UK context, social clubs tend to be venues where people who share a common interest (e.g. music, politics, charity work, sport) congregate and socialize.
3. Respondents' occupations feature after direct quotations to indicate socioeconomic background.
4. Pseudonyms are used throughout this chapter to protect the identity of respondents.
5. This refers to the respondent's gender, age, occupation and method of data collection. E refers to email interview; F refers to face-to-face interview.
6. Rhythmical moving of the head up and down in dancing to heavy metal or rock music (http://www.yourdictionary.com/headbanging).
7. An area in front of a concert stage in which audience members mosh. The act of moshing is to knock against (someone) intentionally while dancing at a rock concert (*The American Heritage Dictionary*, 2000). For a more detailed account of moshing, see Tsitsos (1999).
8. Directed by Sam Mendes, see http://www.imdb.com/title/tt0169547 for more information.
9. Using gymnasium equipment and fitness techniques to improve physical appearance.

Chapter 7—Dance Parties, Lifestyle and Strategies for Ageing

1. Officially referred to as MDMA (Methylenedioxymethamphetamine), the drug quickly acquired the street name of ecstasy, or E.

Chapter 8—Punk, Ageing and the Expectations of Adult Life

1. Short for 'fanzines', these are traditionally free or at-cost amateur publications produced by 'fans'. The zines I used are more professional, but still arguably outside of mainstream journalism.
2. Thanks to Jon Cruz for suggesting this term.

Chapter 9—Alternative Women Adjusting to Ageing, or How to Stay Freaky at 50

1. The emoticon is used in email or text exchanges to denote emotions. In this case, Bee was indicating that she was joking.
2. Bingo wings is slang for the build-up of fat and/or extra skin that hangs from the underside of the upper arms, particularly on women as they age.

Chapter 10—The Collective Ageing of a Goth Festival

1. It is perhaps worth noting that part of the appeal and success of the Whitby Gothic Weekend probably relates to the town's connection with broader Gothic culture, through its connections with Bram Stoker and its role as the setting for part of his famous novel, *Dracula*.
2. This churchyard—St Mary's—is situated high above the town's harbour and formed the setting for one of the scenes in Bram Stoker's *Dracula*.

Chapter 11—'Strong Riot Women' and the Continuity of Feminist Subcultural Participation

1. Usually guitar, bass or drums, but some camps offer instruction on keyboard as well.
2. While rock camps are 'girls only' spaces, many camps do include transgender campers and trans volunteers.
3. We refer to both camps using pseudonyms in this chapter to protect the anonymity of respondents.
4. Not all participants at girls' rock camps identify as Riot Grrrls (or even as feminist), or understand the camp as such; however, many of the central organizers and volunteers who were interviewed did express this connection.
5. Illustrating this point, 'Is that volunteer a boy?' is one of the more commonly asked questions from the girl campers.

Chapter 12—Parenthood and the Transfer of Capital in the Northern Soul Scene

1. I am using the term 'parent' as a catch-all to indicate any significant, influential adult connected to the soul child and his or her domestic setting.

2. A total of eighty-one questionnaires were completed between 2005 and 2006, excluding pilot questionnaires. Ten additional participants were contacted for an update to the scene in 2010–11. While the majority of interviews occurred face to face, some were conducted electronically. Interviewees were asked to complete the questionnaire.
3. All names used are pseudonyms.
4. Dave Godin (1936–2004) was a music journalist and a champion of black American soul music. He was the founder of the Tamla Motown Appreciation Society and is credited with coining the term Deep Soul, as well as Northern Soul.
5. The Wigan Casino was arguably the most celebrated and popular of all Northern Soul nightclubs. Located in Wigan, a town in north-west England, the club hosted Northern Soul all-nighters (all-night dance events, from 2.00 a.m. until 8.00 a.m., and later from 12.30 a.m. until 8.00 a.m.) from 1973 until 1981.
6. A well-known Northern Soul song. Frank Wilson (1965) 'Do I Love You (Indeed I Do)', unreleased.
7. This raises interesting questions regarding the definition of 'scene'. There is not the scope to unpick this here, but this notion of scene activity being divided between activities in designated venues while scene-relevant practices occur in other spaces (e.g. the home) is interesting.
8. Interestingly, this demonstrates a blurring of primary and secondary socialization, if the family is a source of primary socialization, and youth culture and its consumption practices and media are sources of secondary socialization.
9. The Internet has had an impact on this. The consumption of music online and membership of virtual music communities have meant that youth musical practice can occur in the home, beyond the parental gaze. Thank you to Andy Bennett for this comment.
10. While the child's bedroom might have its own soundtrack, Northern does appear there, too, as the couple case study suggests, with one daughter taking Frank's CDs to copy.
11. With this in mind, I do not want to paint a picture of perfection, suggesting that shared musical consumption provides a type of familial cement, binding child and parent and creating moments of shared domestic activity and subsequent domestic harmony. But, of course, music has the potential to bring certain people together, families included.
12. There is no easy way of quantifying whether soul children behave on the scene with parents as they would have done if they had found the scene by other means (i.e. via peers), but it is notable that the environment chosen to share cultural experience with a parent should be one symbolically designated as a parent-free, youth-only zone.
13. A recent poll conducted by members of the Soul Source Northern Soul forum concluded that 93 per cent of soulies (148 votes) did not want to see

children attending all-nighters (February 2011). The discussion surrounding this poll indicated that: child attendance at Northern Soul events was notable and growing; there was disagreement among soulies about the presence of children on the scene; and the topic produced an impassioned response. See http://www.soulsource.co.uk.

14. For instance, if a soul child becomes a Northern Soul DJ with a soul parent's vinyl, then this vinyl has greater potential for remaining, or becoming, valuable.
15. For example, one all-dayer I attended had a mum and her son and daughter on the floor from 3.00 p.m. learning how to spread talcum powder and how to dance to 'The Snake', but by 8.00 p.m. they had left, just as the crowds started to enter to fill the ballroom.
16. The possession of scene knowledge enables the soul child to fit in and gain swift acceptance on the scene. Yet the soul children are still performing within the shadows of their parents; the autonomy of which Thornton (1996) spoke is still eluding the soul children unless they place their own interpretation on the performance of Northern Soul. There is, of course, limited scope to do this because Northern Soul has a blueprint of scene practice. However, different types of venues, different clubs and different regions will have a slightly different interpretation of this blueprint; thus, geographically moving to new Northern Soul venues—as Daniel did by attending university—provides a slice of (perceived) autonomous cultural consumption.

References

Andes, L. (1998), 'Growing Up Punk: Meaning and Commitment Careers in a Contemporary Youth Subculture', in J. S. Epstein (ed.), *Youth Culture: Identity in a Postmodern World*, Oxford: Blackwell.

Attwood, F. (ed.) (2009) *Mainstreaming Sex: The Sexualization of Western Culture*. London and New York: I. B. Tauris.

Banes, S. (1984), 'Breaking Changing', *Village Voice*, 12 June.

Banes, S. (1985), 'Breaking', in N. George, S. Banes, S. Flinker and P. Romanowski (eds), *Fresh: Hip Hop Don't Stop*, New York: Random House.

Bayton, M. (1998), *Frock Rock: Women Performing Popular Music*, New York: Oxford University Press.

Beck, U. (1992), *Risk Society: Towards a New Modernity*, London: Sage.

Beck, U. (1994), 'The Reinvention of Politics: Towards a Theory of Reflexive Modernisation', in U. Beck, A. Giddens and S. Lash (eds), *Reflexive Modernisation: Politics, Tradition and Aesthetics in the Modern Social Order*, Cambridge: Polity Press.

Beck, U. and Beck-Gernsheim, E. (1995), *The Normal Chaos of Love*, Cambridge: Polity Press.

Becker, H. (1963), *Outsiders: Studies in the Sociology of Deviance*, New York: Collier-Macmillan.

Bengry-Howell, A. and Morey, Y. (2010), 'All Together at the Same Thing for the Same Reason: A Temporary Escape from Neoliberalism—Intersubjectivity and Sociability in a Music Festival Context', paper presented at Youth 2010: Identities, Transitions, Cultures conference, University of Surrey.

Bennett, A. (2000), *Popular Music and Youth Culture: Music, Identity and Place*, London: Macmillan.

Bennett, A. (2005), *Culture and Everyday Life*, London: Sage.

Bennett, A. (2006), 'Punk's Not Dead: The Continuing Significance of Punk Rock for an Older Generation of Fans', *Sociology*, 40(2): 219–35.

Bennett, A. (2009), '"Heritage Rock": Rock Music, Re-presentation and Heritage Discourse', *Poetics*, 37(5–6): 474–89.

Bennett, A. (2010), 'Popular Music, Cultural Memory and Everyday Aesthetics', in E. de la Fuente and P. Murphy (eds), *Philosophical and Cultural Theories of Music*, Leiden: Brill.

Bennett, A. (2012), *Growing Old Disgracefully? Popular Music, Ageing and Lifestyle*, Philadelphia, PA: Temple University Press.

Bennett, A. and Kahn-Harris, K. (2004), *After Subculture: Critical Studies in Contemporary Youth Culture*, New York: Palgrave Macmillan.

Bennett, A. and Peterson, R. A. (2004), *Music Scenes: Local, Translocal, and Virtual*, Nashville, TN: Vanderbilt University Press.

Bergling, T. (2004), *Reeling in the Years: Gay Men's Perspectives on Age and Ageism*, Binghamton: Harrington Park Press.

Berlant, L. and Warner, M. (1998), 'Sex in Public', *Critical Inquiry*, 24(2): 547–66.

Biggs, S. (2004), 'Age, Gender, Narratives, and Masquerades', *Journal of Aging Studies*, 18: 45–58.

Biggs, S., Phillipson, C., Leach, R. and Money, A.-M. (2007), 'The Mature Imagination and Consumption Strategies: Age and Generation in the Development of a United Kingdom Baby Boomer Identity', *International Journal of Ageing and Later Life*, 2(2): 31–59.

Binnie, J. (2004), *The Globalization of Sexuality*, London: Sage.

Blaikie, A. (1999), *Ageing and Popular Culture*, Cambridge: Cambridge University Press.

Bourdieu, P. (1984), *Distinction: A Social Critique of the Judgement of Taste*, trans. Richard Nice, London: Routledge.

Bradley, D. (1992), *Understanding Rock'n'Roll: Popular Music in Britain 1955–1964*, Buckingham: Open University Press.

Brake, M. (1985), *Comparative Youth Culture*, New York: Routledge.

Brannen, J. and Nilsen, A. (2002), 'Young People's Time Perspectives: From Youth to Adulthood', *Sociology*, 36(3): 513–37.

Brill, D. (2008), *Goth Culture: Gender, Sexuality and Style*, Oxford: Berg.

Brooker, W. (2007), 'A Sort of Homecoming: Fan Viewing and Symbolic Pilgrimage', in J. Gray, C. Sandvoss and C. Lee Harrington (eds), *Fandom: Identities and Communities in a Mediated World*, New York: New York University Press.

Butler, J. (1990), *Gender Trouble: Feminism and the Subversion of Identity*, London: Routledge.

Butler, J. (1993), *Bodies that Matter: On the Discursive Limits of Sex*, New York: Routledge.

Calcutt, A. (1998), *Arrested Development: Pop Culture and the Erosion of Adulthood*, London: Cassell.

Casey, M. (2007), 'The Queer Unwanted and Their Undesirable "Otherness"', in K. Browne, J. Lim and G. Brown (eds), *Geographies of Sexualities*, Aldershot: Ashgate.

Chabris, C. F. and Simons, D. J. (2010), *The Invisible Gorilla: And Other Ways Our Intuitions Deceive Us*, New York: Crown.

Chaney, D. (1995), 'Creating Memories: Some Images of Aging in Mass Tourism', in M. Featherstone and A. Wernick (eds), *Images of Aging: Cultural Representations of Later Life*, London: Routledge.

Chaney, D. (1996), *Lifestyles*, London: Routledge.

Chideya, F. (1992), 'Revolution Girl-style', *Newsweek*, 23 November: 84–6.

Chodorow, N. (1978), *The Reproduction of Mothering*, Berkeley: University of California Press.

Clark, D. (2003), 'The Death and Life of Punk, the Last Subculture', in D. Muggleton and R. Weinzierl (eds), *The Post Subcultures Reader*, Oxford: Berg.

Clarke, J., Hall, S., Jefferson, T. and Roberts, B. (1976), 'Subcultures, Cultures and Class: A Theoretical Overview', in S. Hall and T. Jefferson (eds), *Resistance Through Rituals: Youth Subcultures in Post-War Britain*, London: Hutchinson.

Cohen, S. (1987), *Folk Devils and Moral Panics: The Creation of the Mods and Rockers*, 3rd ed., Oxford: Basil Blackwell.

Cohen, S. (1991), *Rock Culture in Liverpool: Popular Music in the Making*, Oxford: Clarendon Press.

Collins, D. (1999), ' "No Experts: Guaranteed!": Do-It-Yourself Sex Radicalism and the Production of the Lesbian Sex Zine *Brat Attack*', *Signs*, 25: 65–89.

Cometbus, A. (2011), *Cometbus*, 54 (self-published, no address given).

Crossley, N. (2006), *Reflexive Embodiment in Contemporary Society*, Buckingham: Open University Press.

Cruz, J. M. (2003), *Sociological Analysis of Aging: The Gay Male Perspective*, Binghamton: Harrington Park Press.

Cummings, J. (2008), 'Trade Mark Registered: Sponsorship Within the Indie Music Festival Scene', *Continuum*, 22(5): 687–99.

Davies, J. (2006a), 'Growing Up Punk: Negotiating Ageing Identity in a Local Music Scene', *Symbolic Interaction*, 29(1): 63–9.

Dinshaw, C., Edelman, L., Ferguson, R. A., Freccero, C., Freeman, E., Halberstam, J., Jagose, A., Nealon, C. and Hoang, N. T. (2007), 'Theorizing Queer Temporalities. A Roundtable Discussion', *GLQ: A Journal of Lesbian and Gay Studies*, 13(2–3): 177–95.

Dowd, T. J., Liddle, K. and Nelson, J. (2004), 'Music Festivals as Scenes: Examples from Serious Music, Womyn's Music and SkatePunk', in A. Bennett and R. Peterson (eds), *Music Scenes: Local, Translocal and Virtual*, Nashville, TN: Vanderbilt University Press.

Du Bois-Reymond, M. (1998), ' "I Don't Want to Commit Myself Yet": Young People's Life Concepts', *Journal of Youth Studies*, 1(1): 63–79.

Du Bois-Reymond, M. (2009), 'Models of Navigation and Life Management', in A. Furlong (ed.), *Handbook of Youth and Young Adulthood: New Perspectives and Agendas*, London: Routledge.

Duggan, L. (2002), 'The New Homonormativity: The Sexual Politics of Neoliberalism', in R. Castronovo and D. Nelson (eds), *Materializing Democracy: Towards a Revitalized Cultural Politics*, Durham, NC: Duke University Press.

Durkheim, E. (1995), *The Elementary Forms of Religious Life*, New York: The Free Press.

Ebaugh, H.R.F. (1988), *Becoming an Ex: The Process of Role Exit*, Chicago: University of Chicago Press.
Epstein, J. (ed.) (1998), *Youth Culture: Identity in a Postmodern World*. Malden, MA: Blackwell.
Featherstone, M. (1987), 'Lifestyle and Consumer Culture', *Theory, Culture, and Society*, 4: 54–7.
Featherstone, M. and Hepworth, M. (1989), 'Ageing and Old Age: Reflections on the Postmodern Life Course', in B. Bytheway, T. Keil, P. Allatt and A. Bryman (eds), *Becoming and Being Old: Sociological Approaches to Later Life*, London: Sage.
Featherstone, M. and Hepworth, M. (1991), 'The Mask of Ageing and the Postmodern Lifecourse', in M. Featherstone, M. Hepworth and B. S. Turner (eds), *The Body, Social Process and Cultural Theory*, London: Sage.
Fogarty, M. (2010a), 'Learning Hip Hop Dance: Old Music, New Music, and How Music Migrates', in *Crossing Conceptual Boundaries*, London: University of East London.
Fogarty, M. (2010b), 'Preserving Aesthetics While Getting Paid: Careers for Hip Hop Dancers Today', in *What's It Worth? 'Value' and Popular Music*, IASPM-ANZ Conference Proceedings, Auckland, New Zealand.
Fonarow, W. (1997), 'The Spatial Organization of the Indie Music Gig', in K. Gelder and S. Thornton (eds), *The Subcultures Reader*, New York: Routledge.
Foucault, M. (1976), *The History of Sexuality: Vol. 1—an Introduction*, trans. R. Hurley, London: Penguin.
Frith, S. (1983), *Sound Effects: Youth, Leisure and the Politics of Rock'n'Roll*, London: Constable.
Frith, S. (1987), 'Towards an Aesthetic of Popular Music', in R. Leppert and S. McClary (eds), *Music and Society: The Politics of Composition, Performance and Reception*, Cambridge: Cambridge University Press.
Furlong, A. (ed.) (2009) *Handbook of Youth and Young Adulthood*, London: Routledge.
Gergen, K. J. (1991), *The Saturated Self: Dilemmas of Identity in Contemporary Life*, New York: Basic Books.
Gibson, L. (2009), 'Popular Music and the Life Course: Cultural Commitment, Lifestyles and Identities', unpublished PhD thesis, University of Manchester.
Giddens, A. (1991), *Modernity and Self-Identity: Self and Society in the Late Modern Age*, Cambridge: Polity Press.
Giddens, A. (1994), 'Living in a Post-Traditional Society', in U. Beck, A. Giddens and S. Lash (eds), *Reflexive Modernization: Politics, Tradition and Aesthetics in the Modern Social Order*, Stanford, CA: Stanford University Press.
Gilleard, C. and Higgs, P. (1996), 'Cultures of Ageing: Self, Citizen and the Body', in V. Minichiello (ed.), *Sociology of Aging: International Perspectives*, Melbourne: International Sociological Association, Research Committee on Aging.

Gilleard, C. and Higgs, P. (2002), 'The Third Age: Class, Cohort and Generation', *Ageing and Society*, 22: 369–82.

Gilligan, C. and Brown, L. M. (1992), *Meeting at the Crossroads: Women's Psychology and Girls' Development*, Cambridge, MA: Harvard University Press.

Gilroy, S. (1999), 'Intra-household Power Relations and Their Impact on Women's Leisure', in L. McKie, S. Bowlby and S. Gregory (eds), *Gender, Power and the Household*, London: Macmillan.

Goffman, E. (1963), *Stigma: Notes on the Management of Spoiled Identity*, Englewood Cliffs, NJ: Prentice-Hall.

Goffman, E. (1971 [1959]), *The Presentation of Self in Everyday Life*, Middlesex: Penguin.

Gregory, J. (2007), 'Dancing Politics: Connecting Women's Experiences of Rave in Toronto to Ageism and Patriarchy', unpublished MA thesis, Brock University.

Gregory, J. (2009), 'Too Young to Drink, Too Old to Dance: The Influences of Age and Gender on (Non) Rave Participation', *Dancecult: Journal of Electronic Dance Music Culture*, 1(1): 65–80.

Gregory, J. (2010), '(M)others in Altered States: Prenatal Drug-Use, Risk, Choice, and Responsible Self-Governance', *Social & Legal Studies*, 19: 49–66.

Gregory, J. (2012), '"I've Spent My Whole Life Preparing to Raise Teenagers": Links Between Mothering and Raving', in E. Podnieks (ed.), *Mediating Moms: Mothering in Popular Culture*, Montréal: McGill-Queen's University Press.

Haenfler, R. (2006), *Straight Edge: Clean-Living Youth, Hardcore Punk, and Social Change*, New Brunswick, NH: Rutgers University Press.

Hager, S. (1984), *Hip-hop: The Illustrated History of Rap Music, Break Dancing and Subway Graffiti*, New York: St Martin's Press.

Halberstam, J. (2005), *In a Queer Time and Place: Transgender Bodies, Subcultural Lives*, New York: New York University Press.

Hall, S. and Jefferson, T. (eds) (1976), *Resistance Through Rituals: Youth Subcultures in Post-War Britain*, London: Hutchinson.

Hamilton, A. (2008), 'He's 65—but Mick Jagger Still Looks as if Time is on His Side (Not to Mention the £225m)', *The Times*, n.d., http://entertainment.timesonline.co.uk/tol/arts_and_entertainment/music/article4402485.ece. Accessed 22 October 2011.

Hanisch, C. (1971), 'The Personal is Political', in J. Agel (ed.), *The Radical Therapist: The Radical Therapist Collective*, Oxford: Ballantine Books.

Harrington, C. L. and Bielby, D. D. (2010), 'A Life Course Perspective on Fandom', *International Journal of Cultural Studies*, 13: 429–50.

Hathaway, A. (2001), 'Shortcomings of Harm Reduction: Toward a Morally Invested Drug Reform Strategy', *International Journal of Drug Policy*, 12: 125–37.

Hebdige, D. (1979), *Subculture: The Meaning of Style*, London: Methuen.

Hennion, A. (2007), 'Those Things That Hold Us Together: Taste and Sociology', *Cultural Sociology*, 1(1): 97–114.

Henry, T. (1989), *Break All Rules! Punk Rock and the Making of a Style*, Ann Arbor, MI: UMI Research Press.

Herdt, G. and de Vries, B. (eds) (2003), *Gay and Lesbian Aging*, New York: Springer.

Hills, M. (2002), *Fan Cultures*, London: Routledge.

Hine, C. (ed.) (2005), *Virtual Methods: Issues in Social Research on the Internet*, Oxford: Berg.

Hodkinson, P. (2002), *Goth: Identity, Style and Subculture*, Oxford: Berg.

Hodkinson, P. (2011), 'Ageing in a Spectacular "Youth Culture": Continuities, Changes and Community Among Older Goths', *British Journal of Sociology*, 62(2): 262–82.

Holland, S. (2004), *Alternative Femininities: Body, Age and Identity*, Oxford: Berg.

Holland, S. (2009), 'Preparation and Determination: 3 Vignettes of Gendered Leisure', *Journal of Gender Studies*, 18(1): 35–45.

Holland, S. (ed.) (2008), *Remote Relationships in a Small World*, New York: Peter Lang.

Hunt, C. (2002), 'For Dancers Only: The Story of Wigan Casino', *Mojo Collections*, http://www.chrishunt.biz/features05.html. Accessed 22 October 2011.

Hunt, G. P. and Evans, K. (2008), ' "The Great Unmentionable": Exploring the Pleasures and Benefits of Ecstasy from the Perspectives of Drug-users', *Drugs: Education, Prevention and Policy*, 15: 329–49.

Hunt, S. (2005) *The Life Course: A Sociological Introduction*, Basingstoke: Palgrave.

Hutson, Scott. (2000), 'The Rave: Spiritual Healing in Modern Western Subcultures', *Anthropological Quarterly*, 73: 35–49.

Hutton, F. (2006), *Risky Pleasures? Club Cultures and Feminine Identities*, New Brunswick, NH: Rutgers University Press.

Irwin, J. (1977), *Scenes*, Beverly Hills: Sage.

Juno, A. (1996), *Angry Women in Rock*, New York: Juno Books.

Katz, S. (1996), *Disciplining Old Age: The Formation of Gereontological Knowledge*, Charlottesville: University of Virginia Press.

Kearney, M. C. (1998), 'The Missing Links: Riot Grrrl, Feminism, and Lesbian Culture', in S. Whitely (ed.), *Sexing the Groove: Popular Music and Gender*, London: Routledge.

Kearney, M. C. (2006), *Girls Make Media*, New York: Routledge.

Keightley, K. (2001), 'Reconsidering Rock', in S. Frith, W. Straw and J. Street (eds), *The Cambridge Companion to Pop and Rock*, Cambridge: Cambridge University Press.

Kiley, D. (1995), 'Coming Over All Queer: Theory, Ageing and Embodied Problematics', *Antithesis*, 7(1): 73–103.

Klein, M. (1997), 'Duality and Redefinition: Young Feminism and the Alternative Music Culture', in L. Heywood and J. Drake (eds), *Third Wave Agenda: Being Feminist, Doing Feminism*, Minneapolis: University of Minnesota Press.

Kotarba, J. A. (2002), 'Baby Boomer Rock'n'Roll Fans and the Becoming of Self', in J. A. Kotarba and J. M. Johnson (eds), *Postmodern Existential Sociology*, Walnut Creek, CA: Alta Mira Press.

Laing, D. (1985), *One Chord Wonders: Power and Meaning in Punk Rock,* Milton Keynes: Open University Press.

Leach, R. M. (2011), 'Ageing & Consumer Culture', in D. Southerton (ed.), *Encyclopedia of Consumer Culture*, Thousand Oaks, CA: Sage.

Leafloor, S. (1988), 'Hip Hop Subcultural Thesis', unpublished dissertation, Carleton University, Ottawa.

LeBlanc, L. (1999), *Pretty in Punk: Girls' Gender Resistance in a Boys' Subculture*, New Brunswick, NJ: Rutgers University Press.

Lewis, L. (ed.) (1992), *The Adoring Audience: Fan Cultures and Popular Media*, London: Routledge.

McCaughan, J., Carlson, R., Falck, R. and Siegal, H. (2005), 'From "Candy Kids" to "Chemi-Kids": A Typology of Young Adults Who Attend Raves in the Midwestern United States', *Substance Use & Misuse*, 40: 1503–23.

McNair, B. (2002) *Striptease Culture: Sex, Media and the Democratization of Desire*. London: Routledge.

McKay, G. (1996), *Senseless Acts of Beauty: Cultures of Resistance Since the Sixties*, London: Verso.

McKay, G. (ed.) (1998), *DiY Culture: Party and Protest in Nineties Britain*, London: Verso.

McRobbie, A. (1991), *Feminism and Youth Culture: From 'Jackie' to 'Just Seventeen'*, Boston: Unwin Hyman.

McRobbie, A. and Garber, J. (1976), 'Girls and Subcultures', in S. Hall and T. Jefferson (eds), *Resistance Through Rituals: Youth Subcultures in Post-War Britain*, London: Routledge.

Maffesoli, M. (1996), *The Time of the Tribes: The Decline of Individualism in Mass Society*, trans. D. Smith, London: Sage.

Malbon, B. (1998), 'Clubbing: Consumption, Identity and the Spatial Practices of Every-Night Life', in T. Skelton and G. Valentine (eds), *Cool Places: Geographies of Youth Cultures*, London: Routledge.

Mann, C. and Stewart, F. (2001), 'Internet Interviewing', in J. F. Gubrium and J. A. Holstein (eds), *Handbook of Interview Research: Context and Method*, London: Sage.

Mattson, K. (2001), 'Did Punk Matter? Analyzing the Practices of a Youth Subculture During the 1980s', *American Studies*, 42: 69–97.

Matza, D. and Sykes, G. M. (1961), 'Juvenile Delinquency and Subterranean Values', *American Sociological Review*, 26: 712–19.

Maximum Rock n Roll (1992), 'Punks at 30+, 40+, 50+ Still Giving a Shit!!!', n.d.

Melechi, A. (1993), 'The Ecstasy of Disappearance', in S. Redhead (ed.), *Rave Off: Politics and Deviance in Contemporary Youth Culture*, Aldershot: Avebury.

Meno, J. (2004), Interview with Milo Aukerman, *Punk Planet*, 61: 74–9.
Miles, S. (2000), *Youth Lifestyles in a Changing World*, Buckingham: Open University Press.
Muggleton, D. (2000), *Inside Subculture: The Postmodern Meaning of Style*, Oxford: Berg.
Muggleton, D. and Weinzierl, R. (eds) (2003), *The Post-subcultures Reader*, Oxford: Berg.
Oberg, P. and Tornstam, L. (1999), 'Body Images Among Men and Women of Different Ages', *Ageing and Society*, 19, 629–44.
Ogden, J. and Steward, J. (2000), 'The Role of the Mother–Daughter Relationship in Explaining Weight Concern', *International Journal of Eating Disorders*, 28: 78–83.
Osgerby, B. (2001), *Playboys in Paradise: Masculinity, Youth and Leisure-Style in Modern America*, Oxford: Berg.
Pearsall, M. (ed.) (1997), *The Other Within Us: Feminist Explorations of Women and Aging*. Boulder, CO: Westview Press.
Peterson, R. A. (1990), 'Why 1955? Explaining the Advent of Rock Music', *Popular Music*, 9(1): 97–116.
Phillipson, C and Biggs, S. (1998), 'Modernity and Identity: Themes and Perspectives in the Study of Older Adults', *Journal of Aging and Identity*, 3(1): 11–23.
Phillipson, C., Leach, R., Money, A.-M. and Biggs, S. (2008), 'Social and Cultural Constructions of Ageing: The Case of the Baby Boomers', *Sociological Research Online*, 13(3), http://www.socresonline.org.uk/13/3/5.html. Accessed 22 October 2011.
Pini, M. (2001), *Club Cultures and Female Subjectivity: The Move from Home to House*, New York: Palgrave.
Pipher, M. (1994), *Reviving Ophelia: Saving the Selves of Adolescent Girls*, New York: Ballantine Books.
Plummer, K. (2005). 'Review of the books *Reeling in the Years: Gay Men's Perspectives on Age and Ageism* by T. Bergling; *Sociological Analysis of Aging: The Gay Male Perspective* by J.M. Cruz; *Gay and Lesbian Aging: Research and Future Directions* by G. Herdt and B. de Vries (eds)', *Ageing and Society*, 25(3): 446–8.
Powell, J. L. and Longino, C. H. Jr (2001), 'Towards the Postmodernization of Aging: The Body and Social Theory', *Journal of Aging and Identity*, 6(4): 199–207.
Price, E. (2007), 'Ageing Against the Grain: Gay Men and Lesbians', in P. Burke and J. Parker (eds), *Social Work and Disadvantage: Addressing the Roots of Stigma Through Association*, London: Jessica Kingsley.
Ray, R. E. (2002), 'The Uninvited Guest: Mother/Daughter Conflict in Feminist Gerontology', *Journal of Aging Studies*, 17: 113–28.
Renshaw, S. W. (2002), 'Postmodern Swing Dance and the Presentation of the Unique Self', in A. Kotarba and J. M. Johnson (eds), *Postmodern Existential Sociology*, Walnut Creek, CA: Alta Mira.

Roszak, T. (1969), *The Making of a Counter Culture: Reflections on the Technocratic Society and Its Youthful Opposition*, London: Faber and Faber.

Rubin, G. (1984), 'Thinking Sex: Notes for a Radical Theory of the Politics of Sexuality', in C. Vance (ed.), *Pleasure and Danger: Exploring Female Sexuality*, Boston: Routledge.

Ryan, K. (2004), Interview with Sleater-Kinney, *Punk Planet*, 61: 36–41.

St John, G. (2009), *Technomad: Global Raving Countercultures*, London: Equinox.

St John, G. (ed.) (2004), *Rave Culture and Religion*, London: Routledge.

Sandberg, L. (2008), 'The Old, the Ugly and the Queer: Thinking Old Age in Relation to Queer Theory', *Graduate Journal of Social Science*, 5(2): 117–39.

Sandvoss, C. (2005), *Fans: The Mirror of Consumption*, Cambridge: Polity Press.

Scaruffi, P. (2002), 'History of Rock Music [Based on the Truth]', http://scaruffi.com/history/index.html. Accessed 22 October 2011.

Schilt, K. (2003a), '"I'll Resist with Every Inch and Every Breath": Girls and Zine Making as a Form of Resistance', *Youth and Society*, 35(1): 71–97.

Schilt, K. (2003b), '"A Little Too Ironic": The Appropriation and Packaging of Radical Feminism by the New Angry Women in Rock', *Popular Music and Society*, 26(1): 5–19.

Schilt, K. and Zobl, E. (2008), 'Connecting the Dots: Riot Grrrls, Ladyfests, and the International Grrrl Zine Network', in A. Harris (ed.), *Next Wave Cultures: Feminism, Subcultures, and Activism*, New York: Routledge.

Shively, C. (1980), 'Old and Gay', in P. Mitchell (ed.), *Pink Triangles: Radical Perspective of Gay Liberation*, Boston: Alyson.

Simpson, M. (2009), 'Georgie Boys', *Out*, n.d., http://out.com/detail.asp?id=24502. Accessed 22 October 2011.

Sinker, D. (2003), Interview with Ariel Gore, *Punk Planet*, 55: 60–3.

Sinker, D. (2004), Interview with Ian MacKaye, *Punk Planet*, 61: 30–5.

Skelton, T. and Valentine, G. (eds) (1998), *Cool Places: Geographies of Youth Cultures*, New York: Routledge.

Smith, N. (2009), 'Beyond the Master Narrative of Youth: Researching Ageing Popular Music Scenes', in D. B. Scott (ed.), *The Ashgate Research Companion to Popular Musicology*, Aldershot: Ashgate.

Smith, P. (2004), *Two of Us: The Story of a Father, a Son and the Beatles*, Boston: Houghton Mifflin.

Spencer, L. (1993), 'Grrrls Only', *Washington Post*, 3 January: C1–C2.

Straw, W. (1991), 'Systems of Articulation, Logics of Change: Communities and Scenes in Popular Music', *Cultural Studies*, 5(3): 368–88.

Sullivan, N. (2003), *A Critical Introduction to Queer Theory*, Armadale, Vic: Circa Books.

Tanner, J. and Arnett, J. (2009), 'The Emergence of "Emerging Adulthood": The New Life Stage Between Adolescence and Young Adulthood', in A. Furlong (ed.), *Handbook of Youth and Young Adulthood*, London: Routledge.

Taylor, J. (2010), 'Queer Temporalities: The Significance of "Music Scene" Participation in the Social Identities of Middle-aged Queers', *Sociology*, 44(5): 893–907.

Taylor, J. (2011), 'The Intimate Insider: Negotiating the Ethics of Friendship when Doing Insider Research', *Qualitative Research*, 11(1): 3–22.

Taylor, Y. (2008), '"That's Not Really My Scene": Working-class Lesbians in (and Out of) Place', *sexualities*, 11(5): 523–46.

Thébaud, S. (2010), 'Masculinity, Bargaining, and Breadwinning: Understanding Men's Housework in the Cultural Context of Paid Work', *Gender & Society* 24(3): 330–54.

Thomas, H. (2003), *The Body, Dance and Cultural Theory*, New York: Palgrave.

Thompson, P., Itzin, C. and Abendstern, M. (1990), *I Don't Feel Old: The Experience of Later Life*, Oxford: Oxford University Press.

Thomsson, H. (1999), 'Yes, I Used to Exercise But . . . a Feminist Study of Exercise in the Life of Swedish Women', *Journal of Leisure Research*, 31(1): 35–56.

Thornton, S. (1995), *Club Cultures: Music, Media and Subcultural Capital*, Cambridge: Polity Press.

Torkelson, J. (2010), 'Life After (Straightedge) Subculture', *Qualitative Sociology*, 33(3): 257–74.

Tsitsos, W. (1999), 'Rules of Rebellion: Slamdancing, Moshing, and the American Alternative Scene', *Popular Music*, 18: 397–414.

Turner, B. S. (1995), 'Ageing and Identity: Some Reflections on the Somatization of the Self', in M. Featherstone and A. Wernick (eds), *Images of Ageing: Cultural Representations of Later Life*, London: Routledge, 245–62.

Turner, B. S. (2001), 'Disability and the Sociology of the Body', in G. L. Albrecht, K. D. Seelman and M. Bury (eds), *Handbook of Disability Studies*, Thousand Oaks, CA: Sage.

Twigg, J. (2007), 'Clothing, Age and the Body: A Critical Review', *Ageing and Society*, 27: 285–305.

Usmiani, S. and Daniluk, J. (1997) 'Mothers and Their Adolescent Daughters: Relationship Between Self-esteem, Gender Role Identity and Body Image', *Journal of Youth and Adolescence*, 26(1): 45–62.

Vroomen, L. (2004), 'Kate Bush: Teen Pop and Older Female Fans', in A. Bennett and R. A. Peterson (eds), *Music Scenes: Local, Translocal and Virtual*, Nashville, TN: Vanderbilt University Press.

Warner, M. (1993), 'Introduction', in M. Warner (ed.), *Fear of a Queer Planet: Queer Politics and Social Theory*, Minneapolis: University of Minnesota Press.

Weinstein, D. (2000), *Heavy Metal: The Music and Its Culture*, 2nd ed., New York: Da Capo.

Whyte, W. F. (1943), *Street Corner Society: The Social Structure of an Italian Slum*, Chicago: Chicago University Press.

Wilkins, A. C. (2004), 'So Full of Myself as a Chick': Goth Women, Sexual Independence, and Gender Egalitarianism, *Gender & Society*, 18: 328–49.

Williams, J. P. (2011), *Subcultural Theory: Traditions and Concepts*, Cambridge: Berg.

Williams, J. P. and Copes, H. (2005), ' "How Edge are You?" Constructing Authentic Identities and Subcultural Boundaries in a Straightedge Internet Forum', *Symbolic Interaction*, 28(1): 67–89.

Willis, P. E. (1978), *Profane Culture*, Boston: Routledge.

Wood, R. T. (2006), *Straightedge: Complexity and Contradictions of a Subculture*, Syracuse, NY: Syracuse University Press.

Woodward, K. (2006), 'Performing Age, Performing Gender', *National Women's Studies Association Journal*, 18: 162–89.

Ziegler, C. (2004), Interview with Mike Watt, *Punk Planet*, 61: 42–7.

Index